Introduction to
INTERVAL ANALYSIS

Introduction to
INTERVAL ANALYSIS

Ramon E. Moore
Worthington, Ohio

R. Baker Kearfott
University of Louisiana at Lafayette
Lafayette, Louisiana

Michael J. Cloud
Lawrence Technological University
Southfield, Michigan

siam.
Society for Industrial and Applied Mathematics
Philadelphia

10 9 8 7 6 5 4 3 2 1

Library of Congress Cataloging-in-Publication Data

Moore, Ramon E.
Introduction to interval analysis / Ramon E. Moore, R. Baker Kearfott, Michael J. Cloud.
p. cm.
Includes bibliographical references and index.
ISBN 978-0-898716-69-6
1. Interval analysis (Mathematics) I. Kearfott, R. Baker. II. Cloud, Michael J. III. Title.
QA297.75.M656 2009
511'.42—dc22

2008042348

Contents

Preface

This book is intended primarily for those not yet familiar with methods for computing with intervals of real numbers and what can be done with these methods.

Using a pair $[a, b]$ of computer numbers to represent an interval of real numbers $a \leq x \leq b$, we define an arithmetic for intervals and interval valued extensions of functions commonly used in computing. In this way, an interval $[a, b]$ has a dual nature. It is a new kind of number pair, and it represents a *set* $[a, b] = \{x : a \leq x \leq b\}$. We combine set operations on intervals with interval function evaluations to get algorithms for computing enclosures of sets of solutions to computational problems. A procedure known as outward rounding guarantees that these enclosures are rigorous, despite the roundoff errors that are inherent in finite machine arithmetic. With interval computation we can program a computer to find intervals that contain—with absolute certainty—the exact answers to various mathematical problems. In effect, interval analysis allows us to compute with sets on the real line. Interval vectors give us sets in higher-dimensional spaces. Using multinomials with interval coefficients, we can compute with sets in function spaces.

In applications, interval analysis provides rigorous enclosures of solutions to model equations. In this way we can at least know for sure what a mathematical model tells us, and, from that, we might determine whether it adequately represents reality. Without rigorous bounds on computational errors, a comparison of numerical results with physical measurements does not tell us how realistic a mathematical model is.

Methods of computational error control, based on order estimates for approximation errors, are not rigorous—nor do they take into account rounding error accumulation. Linear sensitivity analysis is not a rigorous way to determine the effects of uncertainty in initial parameters. Nor are Monte Carlo methods, based on repetitive computation, sampling *assumed* density distributions for uncertain inputs. We will not go into interval statistics here or into the use of interval arithmetic in fuzzy set theory.

By contrast, interval algorithms are designed to automatically provide rigorous bounds on accumulated rounding errors, approximation errors, and propagated uncertainties in initial data during the course of the computation.

Practical application areas include chemical and structural engineering, economics, control circuitry design, beam physics, global optimization, constraint satisfaction, asteroid orbits, robotics, signal processing, computer graphics, and behavioral ecology.

Interval analysis has been used in rigorous computer-assisted proofs, for example, Hales' proof of the Kepler conjecture.

An interval Newton method has been developed for solving systems of nonlinear equations. While inheriting the local quadratic convergence properties of the ordinary Newton method, the interval Newton method can be used in an algorithm that is mathematically guaranteed to find all roots within a given starting interval.

Interval analysis permits us to compute interval enclosures for the exact values of integrals. Interval methods can bound the solutions of linear systems with inexact data. There are rigorous interval branch-and-bound methods for global optimization, constraint satisfaction, and parameter estimation problems.

The book opens with a brief chapter intended to get the reader into a proper mindset for learning interval analysis. Hence its main purpose is to provide a bit of motivation and perspective. Chapter 2 introduces the interval number system and defines the set operations (intersection and union) and arithmetic operations (addition, subtraction, multiplication, and division) needed to work within this system.

The first applications of interval arithmetic appear in Chapter 3. Here we introduce outward rounding and demonstrate how interval computation can automatically handle the propagation of uncertainties all the way through a lengthy numerical calculation. We also introduce INTLAB, a powerful and flexible MATLAB toolbox capable of performing interval calculations.

In Chapter 4, some further properties of interval arithmetic are covered. Here the reader becomes aware that not all the familiar algebraic properties of real arithmetic carry over to interval arithmetic. Interval functions—residing at the heart of interval analysis—are introduced in Chapter 5. Chapter 6 deals with sequences of intervals and interval functions, material needed as preparation for the iterative methods to be treated in Chapter 7 (on matrices) and Chapter 8 (on root finding). Chapter 9 is devoted to integration of interval functions, with an introduction to automatic differentiation, an important tool in its own right. Chapter 10 treats integral and differential equations. Finally, Chapter 11 introduces an array of applications including several of those (optimization, etc.) mentioned above.

Various appendices serve to round out the book. Appendix A offers a brief review of set and function terminology that may prove useful for students of engineering and the sciences. Appendix B, the quick-reference Formulary, provides a convenient handbook-style listing of major definitions, formulas, and results covered in the text. In Appendix C we include hints and answers for most of the exercises that appear throughout the book. Appendix D discusses Internet resources (such as additional reading material and software packages—most of them freely available for download) relevant to interval computation. Finally, Appendix E offers a list of INTLAB commands.

Research, development, and application of interval methods is now taking place in many countries around the world, especially in Germany, but also in Austria, Belgium, Brazil, Bulgaria, Canada, China, Denmark, Finland, France, Hungary, India, Japan, Mexico, Norway, Poland, Spain, Sweden, Russia, the UK, and the USA. There are published works in many languages. However, our references are largely to those in English and German, with which the authors are most familiar. We cannot provide a comprehensive bibliography of publications, but we have attempted to include at least a sampling of works in a broad range of topics.

The assumed background for the first 10 chapters is basic calculus plus some familiarity with the elements of scientific computing. The application topics of Chapter 11 may require a bit more background, but an attempt has been made to keep much of the presentation accessible to the nonspecialist, including senior undergraduates or beginning graduate students in engineering, the sciences (physical, biological, economic, etc.), and mathematics.

Of the various interval-based software packages that are available, we chose INTLAB for several reasons. It is fully integrated into the interactive, programmable, and highly popular MATLAB system. It is carefully written, with all basic interval computations represented. Finally, both MATLAB and INTLAB code can be written in a fashion that is clear and easy to debug.

We wish to cordially thank George Corliss, Andreas Frommer, and Siegfried Rump, as well as the anonymous reviewers, for their many constructive comments. We owe Siegfried Rump additional thanks for developing INTLAB and granting us permission to use it in this book. Edward Rothwell and Mark Thompson provided useful feedback on the manuscript. We are deeply grateful to the staff of SIAM, including Senior Acquisitions Editor Elizabeth Greenspan, Developmental Editor Sara J. Murphy, Managing Editor Kelly Thomas, Production Manager Donna Witzleben, Production Editor Ann Manning Allen, Copy Editor Susan Fleshman, and Graphic Designer Lois Sellers.

The book is dedicated to our wives: Adena, Ruth, and Beth.

Ramon E. Moore

R. Baker Kearfott

Michael J. Cloud

Chapter 1

Introduction

1.1 Enclosing a Solution

In elementary mathematics, a problem is "solved" when we write down an exact solution. We solve the equation

$$x^2 + x - 6 = 0$$

by factoring and obtaining the roots $x_1 = -3$ and $x_2 = +2$. Few high school algebra teachers would be satisfied with an answer of the form

One root lies between -4 and -2, while the other lies between 1 and 3.

We need not look far, however, to find even elementary problems where answers of precisely this form are appropriate. The quadratic equation

$$x^2 - 2 = 0$$

has the positive solution $\sqrt{2}$. We understand that there is more to this symbol than meets the eye; the number it designates cannot be represented exactly with a finite number of digits. Indeed, the notion of irrational number entails some process of approximation from above and below. Archimedes (287–212 BCE) was able to bracket π by taking a circle and considering inscribed and circumscribed polygons. Increasing the numbers of polygonal sides, he obtained both an increasing sequence of lower bounds and a decreasing sequence of upper bounds for this irrational number.

Exercise 1.1. Carry out the details of Archimedes' method for a square and a hexagon. (*Note*: Hints and answers to many of the exercises can be found in Appendix C.) □

Aside from irrational numbers, many situations involve quantities that are not exactly representable. In machine computation, representable lower and upper bounds are required to describe a solution rigorously. This statement deserves much elaboration; we shall return to it later on.

The need to *enclose* a number also arises in the physical sciences. Since an experimentally measured quantity will be known with only limited accuracy, any calculation involving this quantity must begin with inexact initial data. Newton's law

$$F = ma \tag{1.1}$$

permits us to solve for the acceleration a of a body *exactly* only when the force F and mass m are known exactly (i.e., to unlimited decimal precision). If the latter quantities are known only to lie in certain ranges, say,

$$F_0 - \Delta F \le F \le F_0 + \Delta F \quad \text{and}$$
$$m_0 - \Delta m \le m \le m_0 + \Delta m,$$

then a can only be bounded above and below:

$$a_l \le a \le a_u. \tag{1.2}$$

For a relation as simple as (1.1), it is easy to determine how a_l and a_u depend on F_0, m_0, ΔF, and Δm.

Exercise 1.2. Carry out this derivation to find explicit bounds on a. □

For more complicated relations, however, ordinary algebra can be cumbersome. The techniques of interval analysis will render the computation of bounds routine. In fact, interval computation was designed for machine implementation! Examples involving hand computation will appear throughout the book, but the reader should bear in mind that this is only for learning purposes.

In interval analysis, we phrase inequality statements in terms of *closed intervals* on the real line. We think of an interval as a set of numbers, which we commonly[1] represent as an ordered pair. Instead of (1.2), for instance, we write

$$a \in [a_l, a_u]. \tag{1.3}$$

We call the interval $[a_l, a_u]$ an *enclosure of* a. The use of simple set notation will repay us many times over in the book; the reader can find a review and summary of this notation in Appendix A. Henceforth, we will prefer notation of the form (1.3) to that of (1.2). However, it is important to keep in mind that placing a number within a closed interval is the same as bounding it above and below.

Let us return to our discussion of scientific calculations. We noted above that measurement error can give rise to uncertainty in "initial data" such as F and m in (1.1). The general sense is that we would *like* to know F and m exactly so that we can get a exactly. In other circumstances, however, we might wish to treat F and m as *parameters* and intentionally vary them to see how a varies. Mathematically, this problem is still treated as in Exercise 1.2, but the shift in viewpoint is evident.

We have one more comment before we end this section. The act of merely enclosing a solution might seem rather weak. After all, it fails to yield *the solution* itself. While this is true, the degree of satisfaction involved in enclosing a solution can depend strongly on the *tightness* of the enclosure obtained. The hypothetical math teacher of the first paragraph might be much happier with answers of the form

$$x_1 \in [-3.001, -2.999], \quad x_2 \in [1.999, 2.001].$$

[1]Other representations are discussed in Chapter 3.

In fact, it is worth noting that if we obtain something like

$$x \in [0.66666, 0.66667],$$

then we do know x to four places. Moreover, there are times when we can and should be satisfied with rather loose bounds on a solution. It might be better to know that $y \in [59, 62]$ *rigorously* than to have an "answer" of the form $y \approx 60$ with no idea of how much error might be present. If we can compute an interval $[a, b]$ containing an exact solution x to some problem, then we can take the *midpoint* $m = (a + b)/2$ of the interval as an approximation to x and have $|x - m| \leq w/2$, where $w = b - a$ is the *width* of the interval. Hence we obtain both an approximate solution *and* error bounds on the approximation.

Exercise 1.3. A computation shows that the mass M of a certain body lies in the interval $[3.7, 3.8]$ kg. State an approximation for M along with error bounds on this approximation. □

1.2 Bounding Roundoff Error

The effects of finite number representation are familiar to anyone who has done scientific computing. Rounding error, if it manages to accumulate sufficiently, can destroy a numerical solution.

The folklore surrounding this subject can be misleading. Here we will provide one example of a computation—involving only a small number of arithmetic operations—that already foils a scheme often thought to be adequate for estimating roundoff error. The idea is to perform the same computation twice, using higher-precision arithmetic the second time. The number of figures to which the two results agree is supposed to be the number of correct figures in the first result.

Example 1.1. Consider the recursion formula

$$x_{n+1} = x_n^2 \qquad (n = 0, 1, 2, \ldots), \tag{1.4}$$

and suppose that $x_0 = 1 - 10^{-21}$. We seek x_{75}. Performing the computation with 10-place arithmetic, we obtain the approximate values

$$x_0 = 1, \quad x_1 = 1, \quad \ldots, \quad x_{75} = 1.$$

Using 20-place arithmetic, we obtain the same sequence of values; hence the two values of x_{75} agree to all 10 places carried in the first computation. However, the exact value satisfies $x_{75} < 10^{-10}$. □

Exercise 1.4. Verify this. □

Example 1.1 illustrates that repeating a calculation with higher-precision arithmetic and obtaining the same answer does not show that the answer is correct. The reason was simply that x_1 is not representable exactly in either 10- or 20-place arithmetic. The next example, first given by Rump in [220], shows that the problem can occur in a more subtle way.

Example 1.2. Consider evaluation of f defined by

$$f = 333.75\,b^6 + a^2(11\,a^2b^2 - b^6 - 121\,b^4 - 2) + 5.5\,b^8 + a/(2b)$$
$$\text{with } a = 77617.0 \text{ and } b = 33096.0.$$

Computing powers by successive multiplications on an IBM 370 system using single, double, and extended precision (approximately 7, 16, and 33 decimal digits, respectively), Rump obtained the following results:

$$\begin{array}{lcl}
\text{single precision} & f = & \underline{1.172603}61\ldots \\
\text{double precision} & f = & \underline{1.172603940053178}47\ldots \\
\text{extended precision} & f = & \underline{1.172603940053178}63185\ldots.
\end{array}$$

The underlining indicates agreement in digits from one computation to the next. We might be tempted to conclude that f is close to 1.172603. However, the exact result is $f = -0.827396\ldots.$ □

Exercise 1.5. How many digits of precision are required to find the value of f in Example 1.2 correct to six decimal digits? Can we know when we have these six digits correct? *Preliminary hint*: We will discuss INTLAB in Chapter 3, after explaining machine implementations of interval arithmetic. Example 3.6 gives an INTLAB program that can compute rigorous bounds for f to a specified accuracy. □

These examples make it clear that repeating a calculation with more precision does not necessarily provide a basis for determining the accuracy of the results. In many cases[2] it *is* true that by carrying *enough* places a result of arbitrarily high accuracy can be found in any computation involving only a finite number of real arithmetic operations beginning with exactly known real numbers. However, it is often prohibitively difficult to tell in advance of a computation how many places must be carried to guarantee results of required accuracy.

If instead of simply computing a numerical approximation using limited-precision arithmetic and then worrying later about the accuracy of the results, we proceed in the spirit of the method of Archimedes to construct intervals known in advance to contain the desired exact result, then our main concerns will be the narrowness of the intervals we obtain and the amount of computation required to get them. The methods treated in this book will yield for Example 1.1, for instance, an interval close to [0, 1] using only 10-place *interval arithmetic*. However, they will yield an interval of arbitrarily small width containing the exact result by carrying enough places. In this case, obviously, more than 20 places are needed to avoid getting 1 for the value of x_0.

We have chosen just two examples for illustration. There are many others in which the results of single, double, and quadruple precision arithmetic all agree to the number of places carried but are all wrong—even in the first digit.

[2]There are cases in which no amount of precision can rectify a problem, such as when a final result depends on testing exact equality between the result of a floating point computation and another floating point number. The code "`IF` $\sin(2 \cdot \arccos(0)) == 0$ `THEN` $f = 0$ `ELSE` $f = 1$" should return $f = 0$, but it may always return $f = 1$ regardless of the precision used.

1.3 Number Pair Extensions

From time to time, mathematicians have found it necessary to produce a new number system by extending an old one. Extensions of number systems involving *ordered pairs* of numbers from a given system are commonplace. The rational numbers are essentially ordered pairs of integers m/n. The complex numbers are ordered pairs of real numbers (x, y). In each case, arithmetic operations are defined with rules for computing the components of a pair resulting from an arithmetic operation on a pair of pairs. For example, we use the rule

$$(x_1, y_1) + (x_2, y_2) = (x_1 + x_2, y_1 + y_2)$$

to add complex numbers. Pairs of special form are equivalent to numbers of the original type: for example, each complex number of the form $(x, 0)$ is equivalent to a real number x.

In Chapter 2 we will consider another such extension of the real numbers—this time, to the system of closed intervals.

Chapter 2

The Interval Number System

2.1 Basic Terms and Concepts

Recall that the closed interval denoted by $[a, b]$ is the set of real numbers given by

$$[a, b] = \{x \in \mathbb{R} : a \leq x \leq b\}.$$

Although various other types of intervals (open, half-open) appear throughout mathematics, our work will center primarily on closed intervals. In this book, the term *interval* will mean *closed interval*.

Endpoint Notation, Interval Equality

We will adopt the convention of denoting intervals and their endpoints by capital letters. The left and right endpoints of an interval X will be denoted by $\underline{X}$ and $\overline{X}$, respectively. Thus,

$$X = [\underline{X}, \overline{X}]. \tag{2.1}$$

Two intervals X and Y are said to be *equal* if they are the same sets. Operationally, this happens if their corresponding endpoints are equal:

$$X = Y \quad \text{if} \quad \underline{X} = \underline{Y} \text{ and } \overline{X} = \overline{Y}. \tag{2.2}$$

Degenerate Intervals

We say that X is *degenerate* if $\underline{X} = \overline{X}$. Such an interval contains a single real number x. By convention, we agree to *identify* a degenerate interval $[x, x]$ with the real number x. In this sense, we may write such equations as

$$0 = [0, 0]. \tag{2.3}$$

Intersection, Union, and Interval Hull

The *intersection* of two intervals X and Y is *empty* if either $\overline{Y} < \underline{X}$ or $\overline{X} < \underline{Y}$. In this case we let $\emptyset$ denote the empty set and write

$$X \cap Y = \emptyset,$$

indicating that X and Y have no points in common. Otherwise, we may define the intersection $X \cap Y$ as the interval

$$\begin{aligned} X \cap Y &= \{z : z \in X \text{ and } z \in Y\} \\ &= \left[\max\{\underline{X}, \underline{Y}\},\ \min\{\overline{X}, \overline{Y}\}\right]. \end{aligned} \tag{2.4}$$

In this latter case, the *union* of X and Y is also an interval:

$$\begin{aligned} X \cup Y &= \{z : z \in X \text{ or } z \in Y\} \\ &= \left[\min\{\underline{X}, \underline{Y}\},\ \max\{\overline{X}, \overline{Y}\}\right]. \end{aligned} \tag{2.5}$$

In general, the union of two intervals is not an interval. However, the *interval hull* of two intervals, defined by

$$X \underline{\cup} Y = \left[\min\{\underline{X}, \underline{Y}\},\ \max\{\overline{X}, \overline{Y}\}\right], \tag{2.6}$$

is always an interval and can be used in interval computations. We have

$$X \cup Y \subseteq X \underline{\cup} Y \tag{2.7}$$

for any two intervals X and Y.

Example 2.1. If $X = [-1, 0]$ and $Y = [1, 2]$, then $X \underline{\cup} Y = [-1, 2]$. Although $X \cup Y$ is a disconnected set that cannot be expressed as an interval, relation (2.7) still holds. Information is lost when we replace $X \cup Y$ with $X \underline{\cup} Y$, but $X \underline{\cup} Y$ is easier to work with, and the lost information is sometimes not critical. □

On occasion we wish to save both parts of an interval that gets split into two disjoint intervals. This occurs with the use of the interval Newton method discussed in Chapter 8.

Importance of Intersection

Intersection plays a key role in interval analysis. If we have two intervals containing a result of interest—regardless of how they were obtained—then the intersection, which may be narrower, also contains the result.

Example 2.2. Suppose two people make independent measurements of the same physical quantity q. One finds that $q = 10.3$ with a measurement error less than 0.2. The other finds that $q = 10.4$ with an error less than 0.2. We can represent these measurements as the intervals $X = [10.1, 10.5]$ and $Y = [10.2, 10.6]$, respectively. Since q lies in both, it also lies in $X \cap Y = [10.2, 10.5]$. An empty intersection would imply that at least one of the measurements is wrong. □

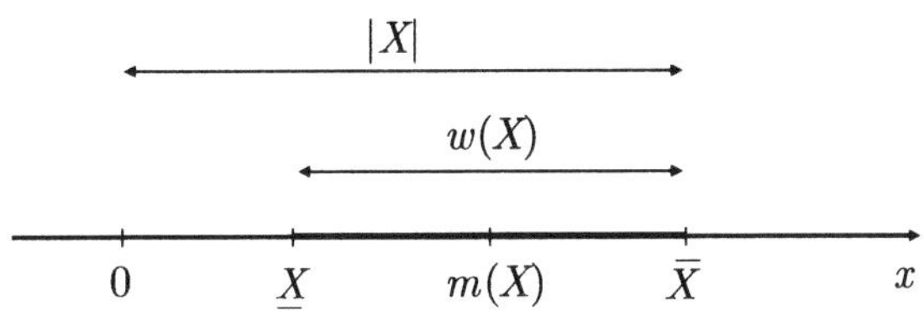

Figure 2.1. *Width, absolute value, and midpoint of an interval.*

Width, Absolute Value, Midpoint

A few other terms will be useful in the book:

1. The *width* of an interval X is defined and denoted by

$$w(X) = \overline{X} - \underline{X}. \tag{2.8}$$

2. The *absolute value* of X, denoted $|X|$, is the maximum of the absolute values of its endpoints:

$$|X| = \max\{|\underline{X}|, |\overline{X}|\}. \tag{2.9}$$

   Note that $|x| \leq |X|$ for every $x \in X$.

3. The *midpoint* of X is given by

$$m(X) = \tfrac{1}{2}(\underline{X} + \overline{X}). \tag{2.10}$$

See Figure 2.1.

Example 2.3. Let $X = [0, 2]$ and $Y = [-1, 1]$. The intersection and union of X and Y are the intervals

$$X \cap Y = [\max\{0, -1\}, \min\{2, 1\}] = [0, 1],$$
$$X \cup Y = [\min\{0, -1\}, \max\{2, 1\}] = [-1, 2].$$

We have $w(X) = w(Y) = 2$ and, for instance,

$$|X| = \max\{0, 2\} = 2.$$

The midpoint of Y is $m(Y) = 0$. □

2.2 Order Relations for Intervals

We know that the real numbers are ordered by the relation $<$. This relation is said to be *transitive*: if $a < b$ and $b < c$, then $a < c$ for any a, b, and $c \in \mathbb{R}$. A corresponding relation can be defined for intervals, and we continue to use the same symbol for it:

$$X < Y \quad \text{means that} \quad \overline{X} < \underline{Y}. \tag{2.11}$$

For instance, $[0, 1] < [2, 3]$, and we still have

$$A < B \text{ and } B < C \implies A < C. \tag{2.12}$$

Recalling the notation of (2.3), we can call X *positive* if $X > 0$ or *negative* if $X < 0$. That is, we have $X > 0$ if $x > 0$ for all $x \in X$.

Another transitive order relation for intervals is set inclusion:

$$X \subseteq Y \quad \text{if and only if} \quad \underline{Y} \le \underline{X} \text{ and } \overline{X} \le \overline{Y}. \tag{2.13}$$

For example, we have $[1, 3] \subseteq [0, 3]$. This is a *partial ordering*: not every pair of intervals is comparable under set inclusion. For example, if X and Y are overlapping intervals such as $X = [2, 5]$ and $Y = [4, 20]$, then X is not contained in Y, nor is Y contained in X. However, $X \cap Y = [4, 5]$, contained in both X and Y.

2.3 Operations of Interval Arithmetic

The notion of the degenerate interval permits us to regard the system of closed intervals as an extension of the real number system. Indeed, there is an obvious one-to-one pairing

$$[x, x] \leftrightarrow x \tag{2.14}$$

between the elements of the two systems. Let us take the next step in regarding an interval as a new type of numerical quantity.

Definitions of the Arithmetic Operations

We are about to define the basic arithmetic operations between intervals. The key point in these definitions is that *computing with intervals is computing with sets*. For example, when we add two intervals, the resulting interval is a set containing the sums of all pairs of numbers, one from each of the two initial sets. By definition then, the *sum* of two intervals X and Y is the set

$$X + Y = \{x + y \colon x \in X,\ y \in Y\}. \tag{2.15}$$

We will return to an *operational* description of addition momentarily (that is, to the task of obtaining a formula by which addition can be easily carried out). But let us define the remaining three arithmetic operations. The *difference* of two intervals X and Y is the set

$$X - Y = \{x - y \colon x \in X,\ y \in Y\}. \tag{2.16}$$

The *product* of X and Y is given by

$$X \cdot Y = \{xy \colon x \in X,\ y \in Y\}. \tag{2.17}$$

We sometimes write $X \cdot Y$ more briefly as XY. Finally, the *quotient* X/Y is defined as

$$X/Y = \{x/y \colon x \in X,\ y \in Y\} \tag{2.18}$$

provided[3] that $0 \notin Y$. Since all these definitions have the same general form, we can summarize them by writing

$$X \odot Y = \{x \odot y : x \in X,\ y \in Y\}, \tag{2.19}$$

where $\odot$ stands for any of the four binary operations introduced above. We could, in fact, go further and define functions of interval variables by treating these, in a similar fashion, as "unary operations." That is, we can define

$$f(X) = \{f(x) : x \in X\}, \tag{2.20}$$

where, say, $f(x) = x^2$ or $f(x) = \sin x$. However, we shall postpone further discussion of interval functions until Chapter 5.

Endpoint Formulas for the Arithmetic Operations

Addition

Let us find an operational way to add intervals. Since

$$x \in X \quad \text{means that} \quad \underline{X} \le x \le \overline{X}$$

and

$$y \in Y \quad \text{means that} \quad \underline{Y} \le y \le \overline{Y},$$

we see by addition of inequalities that the numerical sums $x + y \in X + Y$ must satisfy

$$\underline{X} + \underline{Y} \le x + y \le \overline{X} + \overline{Y}.$$

Hence, the formula

$$X + Y = \left[\underline{X} + \underline{Y},\ \overline{X} + \overline{Y}\right] \tag{2.21}$$

can be used to implement (2.15).

Example 2.4. Let $X = [0, 2]$ and $Y = [-1, 1]$ as in Example 2.3. Then

$$X + Y = [0 + (-1),\ 2 + 1] = [-1, 3].$$

This is not the same as $X \cup Y = [-1, 2]$. □

Exercise 2.1. Find $X + Y$ and $X \cup Y$ if $X = [5, 7]$ and $Y = [-2, 6]$. □

Subtraction

The operational formula (2.21) expresses $X + Y$ conveniently in terms of the endpoints of X and Y. Similar expressions can be derived for the remaining arithmetic operations. For subtraction we add the inequalities

$$\underline{X} \le x \le \overline{X} \quad \text{and} \quad -\overline{Y} \le -y \le -\underline{Y}$$

[3] We remove this restriction with extended arithmetic, described in section 8.1.

to get

$$\underline{X} - \overline{Y} \le x - y \le \overline{X} - \underline{Y}.$$

It follows that

$$X - Y = \left[\underline{X} - \overline{Y}\,,\, \overline{X} - \underline{Y}\right]. \tag{2.22}$$

Note that

$$X - Y = X + (-Y),$$

where

$$-Y = \left[-\overline{Y}\,,\, -\underline{Y}\right] = \{y:\ -y \in Y\}.$$

Observe the reversal of endpoints that occurs when we find the negative of an interval.

Example 2.5. If $X = [-1, 0]$ and $Y = [1, 2]$, then

$$-Y = [-2, -1]$$

and $X - Y = X + (-Y) = [-3, -1]$. □

Exercise 2.2. Find $X - Y$ if $X = [5, 6]$ and $Y = [-2, 4]$. □

Exercise 2.3. Do we have $X - X = 0$ in general? Why or why not? □

Multiplication

In terms of endpoints, the product $X \cdot Y$ of two intervals X and Y is given by

$$X \cdot Y = [\min S\,,\ \max S]\,, \quad \text{where } S = \{\underline{X}\underline{Y}\,,\ \underline{X}\overline{Y}\,,\ \overline{X}\underline{Y}\,,\ \overline{X}\overline{Y}\}. \tag{2.23}$$

Example 2.6. Let $X = [-1, 0]$ and $Y = [1, 2]$. Then

$$S = \{-1 \cdot 1\,,\ -1 \cdot 2\,,\ 0 \cdot 1\,,\ 0 \cdot 2\} = \{-1\,,\ -2\,,\ 0\}$$

and $X \cdot Y = [\min S, \max S] = [-2, 0]$. We also have, for instance, $2Y = [2, 2] \cdot [1, 2] = [2, 4]$. □

The multiplication of intervals is given in terms of the minimum and maximum of four products of endpoints. Actually, by testing for the signs of the endpoints $\underline{X}$, $\overline{X}$, $\underline{Y}$, and $\overline{Y}$, the formula for the endpoints of the interval product can be broken into nine special cases. In eight of these, only two products need be computed. The cases are shown in Table 2.1. Whether this table, equation (2.23), or some other scheme is most efficient in an implementation of interval arithmetic depends on the programming language and the host hardware. Before proceeding further, we briefly mention the wide availability of self-contained interval software. Many programming languages allow interval data types for which all necessary computational details—such as those in Table 2.1—are handled automatically. See Appendix D for further information.

Table 2.1. *Endpoint formulas for interval multiplication.*

Case	$\underline{X \cdot Y}$	$\overline{X \cdot Y}$
$0 \le \underline{X}$ and $0 \le \underline{Y}$	$\underline{X} \cdot \underline{Y}$	$\overline{X} \cdot \overline{Y}$
$\underline{X} < 0 < \overline{X}$ and $0 \le \underline{Y}$	$\underline{X} \cdot \overline{Y}$	$\overline{X} \cdot \overline{Y}$
$\overline{X} \le 0$ and $0 \le \underline{Y}$	$\underline{X} \cdot \overline{Y}$	$\overline{X} \cdot \underline{Y}$
$0 \le \underline{X}$ and $\underline{Y} < 0 < \overline{Y}$	$\overline{X} \cdot \underline{Y}$	$\overline{X} \cdot \overline{Y}$
$\overline{X} \le 0$ and $\underline{Y} < 0 < \overline{Y}$	$\underline{X} \cdot \overline{Y}$	$\underline{X} \cdot \underline{Y}$
$0 \le \underline{X}$ and $\overline{Y} \le 0$	$\overline{X} \cdot \underline{Y}$	$\underline{X} \cdot \overline{Y}$
$\underline{X} < 0 < \overline{X}$ and $\overline{Y} \le 0$	$\overline{X} \cdot \underline{Y}$	$\underline{X} \cdot \underline{Y}$
$\overline{X} \le 0$ and $\overline{Y} \le 0$	$\overline{X} \cdot \overline{Y}$	$\underline{X} \cdot \underline{Y}$
$\underline{X} < 0 < \overline{X}$ and $\underline{Y} < 0 < \overline{Y}$	$\min\{\underline{X}\overline{Y}, \overline{X}\underline{Y}\}$	$\max\{\underline{X}\underline{Y}, \overline{X}\overline{Y}\}$

Division

As with real numbers, division can be accomplished via multiplication by the reciprocal of the second operand. That is, we can implement equation (2.18) using

$$X/Y = X \cdot (1/Y), \tag{2.24}$$

where

$$1/Y = \{y : 1/y \in Y\} = \left[1/\overline{Y},\ 1/\underline{Y}\right]. \tag{2.25}$$

Again, this assumes $0 \notin Y$.

Example 2.7. We can use division to solve the equation $ax = b$, where the coefficients a and b are only known to lie in certain intervals A and B, respectively. We find that x must lie in B/A. However, this is *not* to say that $A \cdot (B/A) = B$. □

Exercise 2.4. Compute the following interval products and quotients:

(a) $[-2, -1] \cdot [-1, 1]$, (b) $[-2, 4] \cdot [-3, 1]$,

(c) $[1, 2]/[-5, -3]$, (d) $[-1, 2]/[5, 7]$. □

A Useful Formula

Any interval X can be expressed as

$$\begin{aligned} X &= m(X) + \left[-\tfrac{1}{2}w(X),\ \tfrac{1}{2}w(X)\right] \\ &= m(X) + \tfrac{1}{2}w(X)[-1, 1]. \end{aligned} \tag{2.26}$$

Example 2.8. If $X = [0, 2]$, then by (2.26) we can write $X = 1 + [-1, 1]$. □

This idea is useful when we employ an interval to describe a quantity in terms of its measured value m and a measurement uncertainty of no more than $\pm w/2$:

$$m \pm \tfrac{w}{2} = \left[m - \tfrac{w}{2}, m + \tfrac{w}{2}\right]. \tag{2.27}$$

2.4 Interval Vectors and Matrices

By an *n-dimensional interval vector*, we mean an ordered n-tuple of intervals

$$(X_1, \ldots, X_n).$$

We will also denote interval vectors by capital letters such as X.

Example 2.9. A two-dimensional interval vector

$$X = (X_1, X_2) = \left(\left[\underline{X}_1, \overline{X}_1\right], \left[\underline{X}_2, \overline{X}_2\right]\right)$$

can be represented as a rectangle in the x_1x_2-plane: it is the set of all points (x_1, x_2) such that

$$\underline{X}_1 \le x_1 \le \overline{X}_1 \quad \text{and} \quad \underline{X}_2 \le x_2 \le \overline{X}_2. \qquad □$$

With suitable modifications, many of the notions for ordinary intervals can be extended to interval vectors.

1. If $x = (x_1, \ldots, x_n)$ is a real vector and $X = (X_1, \ldots, X_n)$ is an interval vector, then we write

$$x \in X \quad \text{if} \quad x_i \in X_i \text{ for } i = 1, \ldots, n.$$

2. The intersection of two interval vectors is empty if the intersection of any of their corresponding components is empty; that is, if $X_i \cap Y_i = \emptyset$ for some i, then $X \cap Y = \emptyset$. Otherwise, for $X = (X_1, \ldots, X_n)$ and $Y = (Y_1, \ldots, Y_n)$ we have

$$X \cap Y = (X_1 \cap Y_1, \ldots, X_n \cap Y_n).$$

This is again an interval vector.

3. If $X = (X_1, \ldots, X_n)$ and $Y = (Y_1, \ldots, Y_n)$ are interval vectors, we have

$$X \subseteq Y \quad \text{if} \quad X_i \subseteq Y_i \text{ for } i = 1, \ldots, n.$$

4. The width of an interval vector $X = (X_1, \ldots, X_n)$ is the largest of the widths of any of its component intervals:

$$w(X) = \max_i w(X_i).$$

5. The midpoint of an interval vector $X = (X_1, \ldots, X_n)$ is

$$m(X) = (m(X_1), \ldots, m(X_n)).$$

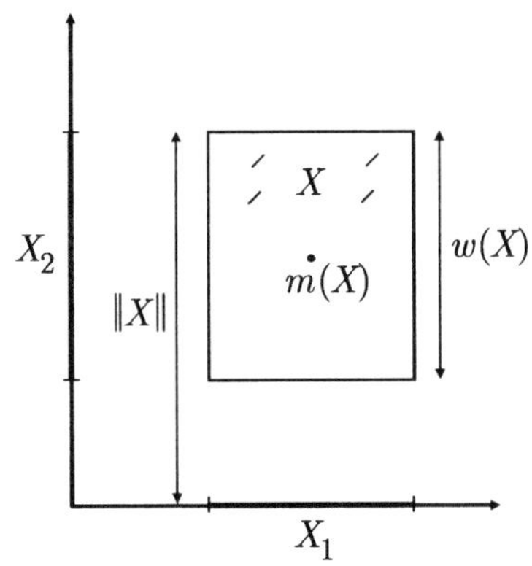

Figure 2.2. *Width, norm, and midpoint of an interval vector* $X = (X_1, X_2)$.

6. The *norm* of an interval vector $X = (X_1, \ldots, X_n)$ is

$$\|X\| = \max_i |X_i|.$$

This serves as a generalization of absolute value.

Example 2.10. Consider the two-dimensional constant interval vector

$$X = (X_1, X_2),$$

where $X_1 = [1, 2]$ and $X_2 = [4, 7]$. We have

$$w(X) = \max\{2 - 1, 7 - 4\} = 3,$$
$$m(X) = \left(\tfrac{1+2}{2}, \tfrac{4+7}{2}\right) = \left(\tfrac{3}{2}, \tfrac{11}{2}\right),$$

and

$$\|X\| = \max\{\max\{|1|, |2|\}, \max\{|4|, |7|\}\} = \max\{2, 7\} = 7.$$

These concepts are illustrated in Figure 2.2. □

Example 2.10 suggests that an interval vector can be thought of as an n-dimensional "box." We will see applications of this idea. Any bounded set of points in n-space can be enclosed by a union of such boxes. Furthermore, we can come arbitrarily close to a given set of points, regardless of its geometric shape. All this leads toward the idea of *computing with sets*, which can be *much* more general and powerful than computing with single points (i.e., with numbers or vectors of numbers).

We can also define an *inner product*

$$P = U_1V_1 + \cdots + U_nV_n$$

between two interval vectors $(U_1, \ldots, U_n)$ and $(V_1, \ldots, V_n)$. The interval P contains all the real numbers defined by values of the inner product of real vectors u and v with real components taken from the given intervals $U_1, \ldots, U_n$ and $V_1, \ldots, V_n$.

Exercise 2.5. A family of Cartesian vectors is given by $(f, 6, -7)$, where $1 \le f \le 3$. Calculate the range of inner products between these vectors and $\frac{1}{\sqrt{5}}(1, 2, 0)$. □

By an *interval matrix* we mean a mean a matrix whose elements are interval numbers. Such matrices are covered further in Chapter 7. In particular, there are some pitfalls, subtleties, and interesting properties of interval matrix-vector multiplication that we will discuss there.

2.5 Some Historical References

Much of the modern literature on interval arithmetic can be traced to R. E. Moore's dissertation [146], through his book [148]. A collection of early papers by Moore and coworkers can be found on the web (see Appendix D for some starting links relevant to the material in this section). The earlier work [238], written independently and often overlooked, also contained many of the ideas expressed in [146]. The paper [238], as well as other early papers dealing with similar concepts, are available on the web. Moore describes the thought process and inspiration that led to his seminal dissertation in [155].

Other researchers began work with interval computations almost contemporary with or shortly after (that is, within a decade of) Moore's early work.

- Eldon Hansen investigated reliable solution of linear systems and other topics; see [58, 59, 60, 61, 62, 67, 68], etc. A decade or so later, he began a lasting collaboration with Bill Walster, including [69, 247]. A notable recent reference is [57].

- William Kahan, known for his work in standardization of floating point arithmetic, had several publications on interval computations, including [87, 88, 89, 90, 91] within several years of Moore's dissertation. Often cited is [89] on extended arithmetic. Kahan devised a closed system in which an interval divided by an interval containing 0 is well defined. *Kahan arithmetic* was originally meant for dealing with continued fractions, a task for which it is particularly suited. It is also consistent with Kahan's philosophy of nonstop exception-free arithmetic, embodied in the IEEE 754 standard for binary floating point arithmetic. In fact, some properties of the binary floating point standard, most notably directed roundings, are important for interval computations; others, such as operations with ∞, facilitate extended arithmetic. The details of both the operations with ∞ in the floating point standard, as well as the details and mathematical underpinnings of extended interval arithmetic, continue to be debated and revised.

- Several researchers at the University of Karlsruhe started a tradition in interval computation that continues to influence mathematics and computer science in Germany:

  - Karl Nickel began at Karlsruhe, where he helped establish the discipline of computer science in Germany. His early publications on interval computations include [19, 41, 175, 176, 177, 178, 179]. Nickel later moved to the University of Freiburg, where he supported the field with work such as editing the *Freiburger Intervallberichte* preprint series, in which there are many gems, some not published elsewhere. One researcher he encouraged at Freiburg is

Arnold Neumaier, presently at the University of Vienna. Neumaier continues to be active in interval computations and global optimization.

- Ulrich Kulisch, starting in 1966, has been influential through his work and that of his students. His early work includes [15, 16, 17, 18, 19, 122, 123], as well as slightly later technical reports at the University of Wisconsin at Madison and the IBM Thomas Watson Research Center. Many of his 49 students hold prominent academic positions in German universities. Two of his early students are Götz Alefeld and Jürgen Herzberger, both receiving the Ph.D. in 1969. Alefeld's early work includes [4, 6, 5]; he has mentored 29 Ph.D. students and continues to hold a chair at the University of Karlsruhe. Herzberger's early work includes his dissertation [73], [74, 75, 76, 77, 78], and a productive collaboration with Alefeld, including [7, 8, 9, 10, 11, 14]. The most famous fruit of this collaboration is the classic introduction to interval analysis [12], which Jon Rokne translated into English a decade later [13]. Another student of Kulisch is Siegfried Rump (1980), presently head of the Institute of Reliable Computing at the Technical University of Hamburg. Rump has done work in error bounds for solutions to linear and nonlinear systems, among other things. He developed the INTLAB toolbox that we use throughout this book.
- Rudolf Krawczyk was at Karlsruhe in 1969 when he published [119], where the much-studied *Krawczyk method* first appeared. (See section 8.2 of this work.)

- Soon after publication of Moore's dissertation, he was invited to give a talk at a seminar on error in digital computing at the Mathematics Research Center (MRC) at the University of Wisconsin. The proceedings of this seminar, published as [195, 196], were edited by Louis Rall. Moore joined MRC and wrote *Interval Analysis* during the summer of 1965. Ulrich Kulisch and Karl Nickel visited the MRC, and the work in interval analysis at the MRC continued through the 1980s, as evidenced in technical reports such as [26, 30, 32, 124, 127, 180, 181, 182, 183, 197, 198, 199, 201, 202, 203, 204, 205, 206, 207, 208, 215, 251, 252, 253, 254, 255].

- There is early work other than at the aforementioned centers. For example, Peter Henrici et al. have studied complex interval arithmetic, such as in [47].

We will draw upon these and more recent references throughout the book.

Chapter 3

First Applications of Interval Arithmetic

3.1 Examples

Almost any scientific computation begins with inexact initial data. Interval arithmetic provides a natural way of incorporating measurement uncertainties directly into a calculation.

Example 3.1. Suppose we wish to calculate the area a of a rectangle from direct measurements of its side lengths l and w. Use of a standard meter stick shows that l equals 1 m to within an uncertainty of 0.0005 m (i.e., to within half the "least count measurement" of 1 mm). Since

$$0.9995 \le l \le 1.0005,$$

we construct an interval $L = [0.9995, 1.0005]$ to represent this side length. Measuring again, we find that $w = 2$ (nominally). Hence, the remaining side should be represented by the interval $W = [1.9995, 2.0005]$. Now

$$\begin{aligned} A &= L \cdot W \\ &= [0.9995, 1.0005] \cdot [1.9995, 2.0005] \\ &= [0.9995 \cdot 1.9995\,,\ 1.0005 \cdot 2.0005] \\ &= [1.99850025\,,\ 2.00150025]\ \text{m}^2, \end{aligned} \tag{3.1}$$

which means, of course, that

$$1.99850025\ \text{m}^2 \le a \le 2.00150025\ \text{m}^2.$$

If we want to know a more accurately, we must measure l and w more accurately. The midpoint of A is 2.00000025. The intervals L, W, and A all carry physical units. □

Let us take this example a bit further. Suppose it is not necessary to know the final answer to eight places. We are therefore tempted to round off the endpoints of A. This is indeed possible, but we must be careful because the true value a of the product lw can fall *anywhere* within the interval that is specified *exactly* by (3.1). Suppose the true value of a is

$$a = 1.99850026\ \text{m}^2. \tag{3.2}$$

Clearly, if we were to round the endpoints of A both *upward* by one digit to obtain

$$A' = [1.9985003\,,\ 2.0015003]\ \text{m}^2,$$

this new interval A' would *not* contain the value (3.2). Such an event would defeat the entire purpose of rigorous computation! Instead, we will *always* implement a procedure called *outward rounding*: we will round in such a way that the left endpoint moves to the left (on the number line) and the right endpoint moves to the right. The resulting interval *contains* the one we started with and hence still has a as a member. In the present example, we could round outwardly to the nearest square millimeter and obtain the interval

$$A'' = [1.998, 2.002]\ \text{m}^2.$$

The statement

$$1.998\ \text{m}^2 \le a \le 2.002\ \text{m}^2$$

is definitely true, and the ability to depend on this is essential as we proceed.

Exercise 3.1. Perform outward rounding, at the third decimal place, on the interval [1.23456, 1.45678]. □

Exercise 3.2. The dimensions of a rectangular box are measured as $w = 7.2 \pm 0.1$, $l = 14.9 \pm 0.1$, and $h = 405.6 \pm 0.2$. Find several intervals containing the volume of the box. □

Exercise 3.3. A Wien bridge electric circuit oscillates at frequency f_0 given by $f_0 = 1/(2\pi RC)$ Hz. If components having nominal values $C = 1$ nF and $R = 15$ kΩ with manufacturing tolerances of 10% are used, in what range must f_0 lie? □

Interval arithmetic also makes it easy to track the propagation of initial uncertainties through a series of calculations.

Example 3.2. Consider the formula

$$V = \sqrt{\frac{2gM}{E(M+E)}} - V_0, \tag{3.3}$$

which has an application discussed in [149]. Suppose we know that

$$\begin{aligned}
g &\in [1.32710, 1.32715](10^{20}),\\
V_0 &\in [2.929, 3.029](10^{4}),\\
M &\in [2.066, 2.493](10^{11}),\\
E &\in [1.470, 1.521](10^{11}),
\end{aligned}$$

and we seek an interval containing V. The necessary calculations are laborious if done by hand, and in Example 3.3 we will show how they can be easily done with INTLAB. It is worth examining the final results at this time, however. We will discuss functions such as square roots systemically in Chapter 5, but for now it will suffice to note that if $x \in [a, b]$ with a and b positive numbers such that $a < b$, then

$$\sqrt{x} \in \left[\sqrt{a}, \sqrt{b}\right].$$

Carrying out the operations indicated in (3.3), we find that

$$V \in [-324, 6394]. \tag{3.4}$$

The interesting part is that instead of (3.3) we can use the expression

$$V = \sqrt{\frac{2g}{E(1+\frac{E}{M})}} - V_0.$$

This is equivalent in ordinary arithmetic, but gives the sharper (i.e., narrower) result

$$V \in [1411.7, 4413.5]. \tag{3.5}$$

So two expressions that are equivalent in ordinary arithmetic can fail to be equivalent in interval arithmetic. This is a very important issue that we will continue to discuss. □

Let us pursue the phenomenon of the previous example using the simpler formula

$$\frac{M}{1+M} = \frac{1}{1+\frac{1}{M}}.$$

For the interval data $M = [14, 15]$, the first expression yields

$$\frac{[14, 15]}{1+[14, 15]} = \frac{[14, 15]}{[15, 16]} = \left[\tfrac{14}{16}, \tfrac{15}{15}\right] = [0.875, 1.0]$$

because we divide by the largest number in the denominator to get the left endpoint of the result. The second expression yields the sharper result

$$\frac{1}{1+\frac{1}{[14,15]}} = \frac{1}{1+\left[\frac{1}{15}, \frac{1}{14}\right]} = \frac{1}{\left[\frac{16}{15}, \frac{15}{14}\right]} = \left[\tfrac{14}{15}, \tfrac{15}{16}\right] \in [0.933, 0.938].$$

The result just shown, namely,

$$\left[\tfrac{14}{15}, \tfrac{15}{16}\right] \in [0.933, 0.938]$$

illustrates how we can maintain rigorous enclosure despite machine rounding error. In decimal form, we would have the *exact* result,

$$\left[\tfrac{14}{15}, \tfrac{15}{16}\right] = [0.933333\ldots, 0.9375].$$

However, the left endpoint is a repeating decimal and cannot be exactly represented by a finite string of digits. Since we can compute only with finite strings of digits, we maintain rigorous enclosure of interval results through outward rounding. Sometimes no rounding is needed, if we have the exact result in no more than the maximum number of digits allowed. If we allow only three places, we get the result shown above. We see that $[0.933, 0.938]$ is the narrowest interval, using only three places, that contains the interval $\left[\frac{14}{15}, \frac{15}{16}\right]$.

3.2 Outwardly Rounded Interval Arithmetic

In practice, outward rounding is implemented at every arithmetic operation—always at the last digit carried. In *optimal outward rounding*, the outwardly rounded left endpoint is the closest machine number less than or equal to the exact left endpoint, and the outwardly rounded right endpoint is the closest machine number greater than or equal to the exact right endpoint. Numerous interval software systems do this automatically (see Appendix D).

Definition 3.1. By *outwardly rounded interval arithmetic (IA)*, we mean interval arithmetic, implemented on a computer, with outward rounding at the last digit carried.

The classic reference on IA is the mathematically rigorous treatment by Kulisch and Miranker [125].

3.3 INTLAB

Software packages that implement interval arithmetic use outward rounding and implement elementary functions of interval arguments (which we will see in Chapter 5). Particularly convenient is INTLAB, available free of charge for noncommercial use. INTLAB provides an interactive environment within MATLAB, a commonly used interactive and programming system both for academic and commercial purposes. One who has MATLAB and has installed INTLAB can easily experiment with interval arithmetic.[4]

Example 3.3. We can use INTLAB to perform the calculations for Example 3.2. At the MATLAB prompt (>>), we begin by issuing the command

```
intvalinit('DisplayInfsup')
```

which directs INTLAB to display intervals using its *infimum-supremum* notation (or lower bound–upper bound representation, i.e., our usual endpoint notation). After providing confirmation[5] of this default display setting, INTLAB returns to the MATLAB prompt and awaits further instruction. We now type

```
>> g = infsup(1.32710e20,1.32715e20)
```

and hit `enter`; INTLAB responds (MATLAB style) with

```
intval g = 1.0e+020 * [ 1.3271 , 1.3272 ]
```

which merely confirms the value of g that we entered. We can suppress this echoing feature by appending a semicolon to the end of our variable assignment:

```
g = infsup(1.32710e20,1.32715e20);
```

Similarly, we can continue to enter the remaining values needed in Example 3.2. Our interactive session appears on screen as

[4]Many books on MATLAB are available (e.g., [79, 145, 214]). The interested reader may wish to consult one before attempting to read further.

[5]For brevity we omit this confirmation from our examples. MATLAB output has been lightly edited to conserve space.

```
>> V0 = infsup(2.929e4,3.029e4);
>> M = infsup(2.066e11,2.493e11);
>> E = infsup(1.470e11,1.521e11);
```

Having entered g, V_0, M, and E, we can use INTLAB to calculate expressions given in terms of these. To request the value of $2gM$, we issue the command

```
2 * g * M
```

and get back

```
intval ans = 1.0e+031 * [ 5.4835 , 6.6172 ]
```

Now we ask INTLAB for $M + E$ and $E(M + E)$:

```
>> M + E
intval ans = 1.0e+011 * [ 3.5360 , 4.0140 ]
>> E * (M+E)
intval ans = 1.0e+022 * [ 5.1979 , 6.1053 ]
```

We continue, deciding to assign a variable name `wide_result`:

```
>> wide_result = 2*g*M / (E*(M+E))
intval wide_result = 1.0e+009 * [ 0.8981 , 1.2731 ]
```

Corresponding to equation (3.4), we compute `wide_V`:

```
>> wide_V = sqrt(wide_result) - V0
intval wide_V = 1.0e+003 * [ -0.3206 , 6.3898 ]
```

Finally, we assign the name `narrow_V` to the sharper result of equation (3.5):

```
>> narrow_V = sqrt(2*g/(E*(1+E/M))) - V0
intval narrow_V = 1.0e+003 * [ 1.4130 , 4.4128 ]
```

To see more digits of this answer, we can issue the command

```
format long
```

and then ask for the value of `narrow_V` again:

```
>> narrow_V
intval narrow_V = 1.0e+003 * [ 1.41309671989000,4.41275778813371 ]
```

INTLAB is a powerful and convenient numerical tool. We will use it freely in subsequent examples. The reader is encouraged to apply it to as many of the exercises as possible. □

Exercise 3.4. Rework Exercises 3.2 and 3.3 using INTLAB. □

INTLAB uses MATLAB's arithmetic, which in turn, generally, uses the IEEE 754 binary standard arithmetic defined partially within the computer's hardware (chips) and partially within software. This means that roughly 16 decimal digits (53 binary digits) are carried. Internally, INTLAB rounds out in the last digit; however, when INTLAB displays results, it rounds the internal representations out so that the displayed result contains the actual result. The number of digits displayed in INTLAB corresponds to the "short" format or the "long" format of MATLAB. The difference between the computational results of Example 3.2 and those using INTLAB directly above is due to the fact that in Example 3.2, only a few decimal digits were carried, with rounding out after each operation, whereas in the INTLAB results, approximately 16 decimal digits were carried.

Example 3.4. The following MATLAB/INTLAB computations illustrate the display of intervals in INTLAB.

```
>> format short
>> rx = 2/3
rx = 0.6667
>> format long
>> rx
rx = 0.666666666666667
>> format short
>> intvalinit('DisplayInfsup')
===> Default display of intervals by infimum/supremum
>> x = infsup(2/3,2/3)
intval x = [ 0.6666 , 0.6667 ]
>> format long
>> x
intval x = [ 0.66666666666666 , 0.66666666666667 ]
```

In these computations, when 2/3 is entered into MATLAB, it is converted (presumably[6]) to the closest IEEE double precision binary number to 2/3 and stored in `rx`. When `rx` is printed, this binary number is converted back to a decimal number, according to the format (short or long) in effect when the output is requested. When the INTLAB command `x = midrad(2/3,0)` is issued, an internal representation[7] corresponding to an interval whose lower bound is less than 2/3 (a number that is not exactly representable in any binary format) and whose upper bound is greater than 2/3 is generated. The lower bound and upper bound are very close together in this case, possibly with no binary numbers in the system between. When output in the short format, the decimal representation of the lower bound is produced to be smaller than or equal to the exact lower bound of the stored binary interval, and the decimal representation of the upper bound is greater than or equal to the exact upper bound of the stored binary interval. Thus, the output interval is wider in the short format than in the long format, but, in any case, the output will always contain the exact value 2/3.

INTLAB provides a special syntax for mathematically rigorous enclosures of decimal numbers that are entered. For example, one may obtain a mathematically rigorous enclosure for the interval $[1.32710 \times 10^{20}, 1.32715 \times 10^{20}]$ from Example 3.3 with the following MATLAB dialogue:

```
>> g = hull(intval('1.32710e20'),intval('1.32715e20'))
intval g = 1.0e+020 * [ 1.3271 , 1.3272 ]
```

The following MATLAB m-file[8] could also be used:

```
function [ivl] = rigorinfsup(x,y)
% [ivl] = rigorinfsup(x,y) returns an interval whose internal
% representation contains a rigorous enclosure for the
% decimal strings x and y. It is an error to call this function
```

[6] We have not verified this on all systems for all inputs.

[7] The representation is not necessarily in terms of lower bound and upper bound, but may be in terms of midpoint and radius.

[8] An m-file is essentially a MATLAB program. Appendix D contains instructions on how to obtain the m-files used in this book. We will say more about m-files later.

```
% if the arguments are not character.
if (ischar(x) & ischar(y))
    ilower = intval(x);
    iupper = intval(y);
    ivl = infsup(inf(ilower), sup(iupper));
else
    display('Error in rigorinfsup; one of arguments is not');
    display('a character string.');
    ivl = infsup(-Inf, Inf);
end;
```

Use of `rigorinfsup` is exemplified in the following short MATLAB dialogue:

```
>> g = rigorinfsup('1.32710e20','1.32715e20')
intval g = 1.0e+020 * [ 1.3271 , 1.3272 ]
```

Above, the output of `rigorinfsup` is given in short format. This is the default format when MATLAB starts, unless the `format long` command has previously been issued in the `startup.m` script or elsewhere in the session.

Caution There is a reason for this special syntax with quotation marks.[9] In particular, we should mention a subtle aspect of INTLAB and other languages implementing interval arithmetic in a similar way.[10] The problem is that when one enters a constant in decimal form, such as `1.32710e20`, the system needs to convert this constant into its internal binary format for further use. However, unless the constant is representable as a fraction whose denominator is a power of 2, there is no internal binary number that corresponds exactly to the decimal number that was entered. The conversion is to a binary number that is near to the decimal number but is not guaranteed to be less than or greater than the corresponding decimal number. Thus, issuing the command `g = infsup(1.32710e20,1.32715e20)` in MATLAB results in an internally stored binary interval for `g` that may not contain the decimal interval $[1.32710 \times 10^{20}, 1.32715 \times 10^{20}]$. This discrepancy usually is not noticed but occasionally causes unexpected results. In contrast, when one uses `infsup('1.32710e20','1.32715e20')`, INTLAB provides a conversion with rigorous rounding. For example, the internal representation for `intval(0.1)` or `infsup(0.1,0.1)` does not necessarily include the exact value 0.1, but `intval('0.1')` does.[11]

Three representations for intervals can appear in INTLAB computations: the infimum-supremum representation, which we have already discussed, the *midpoint-radius representation*, mentioned in Example 3.4, and the *significant digits representation*. The midpoint-radius representation is analogous to the specification of tolerances familiar to engineers. For example, if an electrical resistor is given by $R = 10 \pm 1\ \Omega$, then the midpoint-radius representation of the set of values for R would be $\langle 10, 1\rangle$, whereas the infimum-supremum representation of the interval for R would be $[9, 11]$. (In the midpoint-radius representation, the radius is rounded up for display, to ensure that the displayed interval contains

[9] The quotation marks tell MATLAB that the argument is a string of characters, not a usual number. This allows INTLAB to use a special routine to process this string.

[10] With "operator overloading."

[11] `infsup('0.1','0.1')` does not give the expected results in INTLAB version 5.4 or earlier.

the actual interval.) There are certain advantages and disadvantages to the midpoint-radius representation, both in the display of intervals and in internal computations with intervals.

Example 3.5. Suppose an electrical resistor has a nominal value of 100 Ω and a manufacturing tolerance of 10%. INTLAB permits us to enter this information as follows:

```
>> Ohms = 100;
>> R = midrad(Ohms,0.1*Ohms)
intval R = [ 90.0000 , 110.0000 ]
```

Note, however, that `R` is still displayed in the current default mode, which is `Infsup`. The command

```
intvalinit('DisplayMidrad')
```

will change INTLAB to `Midrad` mode. Now we can see

```
>> R
intval R = < 100.0000 , 10.0000 >
```

Note that we can input a "thin" interval by specifying its radius as zero or by simply not specifying the radius:

```
>> thin = midrad(1,0)
intval thin = < 1.0000 , 0.0000 >
```

We can always switch back to `Infsup` mode and see `thin` in that notation:

```
>> intvalinit('DisplayInfsup')
>> thin
intval thin = [ 1.0000 , 1.0000 ]
```

□

The *significant digits* or *uncertainty representation* is appropriate when we are approximating a single real number and wish to easily see how many digits of it are known with mathematical certainty. The interval is displayed as a single real number, and the last digit displayed is correct to within one unit. The following MATLAB dialogue gives some examples:

```
>> intvalinit('Display_')
>> infsup(1.11,1.12)
intval ans = 1.12__
>> infsup(1.1,1.5)
intval ans = 1.____
>> infsup(1.11,1.13)
intval ans = 1.12__
>> infsup(1.1111111,1.1111112)
intval ans = 1.1111
>> format long
>> infsup(1.11,1.12)
intval ans = 1.12_____________
>> infsup(1.11111,1.11113)
intval ans = 1.11112__________
>> infsup(2,4)
intval ans = 1_.______________
```

In Example 6.4, the INTLAB `Display_` setting will be used to illustrate a convergent interval sequence.

Example 3.6. The following INTLAB program uses the long precision toolbox that comes with INTLAB to find an interval enclosure for f in Example 1.2:

```
function [Intf]= Rump_example(ndigits)
% This evaluates f for Example 1.2 (Rump's counterexample)
% ndigits is the number of digits precision to be used.
% The output Intf is an interval obtained using that many
% digits of precision.
longinit('WithErrorTerm');
longprecision(ndigits);
a = long(77617.0);
b = long(33096.0);
b2 = b*b;
b4 = b2*b2;
b6 = b2*b4;
b8 = b4*b4;
a2=a*a;
f = long(333.75)* b6 + a2*(long(11)* a2*b2 - b6...
    - long(121)*b4 - 2) + long(5.5)*b8 + a/(long(2)* b);
Intf = long2intval(f);
end
```

This is the contents of a MATLAB m-file called `Rump_example.m`. With this file in MATLAB's search path, one can carry on the following MATLAB dialogue:

```
>> Rump_example(10)
intval ans = 1.0e+014 * [ -6.72351402328065 , 5.31613913972737 ]
>> Rump_example(20)
intval ans = 1.0e+014 * [ -6.72351402328065 , 5.31613913972737 ]
>> Rump_example(30)
intval ans = 1.0e+007 * [ -0.04587540000001 , 1.72359720000001 ]
>> Rump_example(100)
intval ans = [ -0.82739605994683 , -0.82739605994682 ]
```

Exercise 3.5. Experiment with `Rump_example` for different values of `ndigits`. What is the minimum number of digits necessary such that the lower and upper bounds of the interval have the same sign? What is the minimum number of digits necessary to ensure that all digits of the answer are correct?

Siegfried Rump, the author of INTLAB, encourages INTLAB users to contribute their own software and extensions.

INTLAB References

In his master's thesis [70], Gareth Hargreaves gives a short tutorial of INTLAB with examples and summarizes many topics covered in this book. He also supplies a number of m-files, partially overlapping with ones we present. Rump describes the INTLAB system in [224], while extensive documentation for INTLAB is integrated into MATLAB's "demo" system, starting with INTLAB version 5.4. Rump describes the way interval arithmetic is implemented within INTLAB in [223]. The INTLAB toolbox features prominently in the solution of 5 of the 10 problems posed in the SIAM 100-digit challenge;[12] this is described in [27].

3.4 Other Systems and Considerations

Shin'ichi Oishi has developed SLAB, a stand-alone MATLAB-like system specifically for interval computations and the automatic result verification it provides. It is available for free (under the GNU license), and has builds for MS-Windows, Linux, and Mac OS. A short description appears in [104].

Tools are available for doing interval computations within traditional programming languages. In Fortran, there are the *ACM Transactions on Mathematical Software* algorithms [101] and [96]. Various groups have also developed tools for use with C++. Prominent among these is C-XSC [80], developed by a group of researchers associated with Ulrich Kulisch at the University of Karlsruhe and presently supported from members of that group at the University of Wuppertal. Extensive libraries have been developed for automatically verified versions of standard computations in numerical analysis using C-XSC; an early version of this toolbox is described in [56], while an update containing more advanced algorithms is [117]. Alternate, widely used class libraries for interval arithmetic in C++ are PROFIL/BIAS [111] and FILIB++ [132]. See Appendix D.

Under certain circumstances, it may be desirable to use higher precision (i.e., more digits) in the interval representations than is available in, say, the IEEE 754 standard for binary floating point arithmetic. Along these lines, Nathalie Revol et al. have developed the multiple precision interval library MPFI [217].

A major thrust by Kulisch and his students at the University of Karlsruhe has been to develop algorithms that provide narrow intervals to rigorously enclose the solutions to point linear systems of equations, even if these linear systems happen to be ill-conditioned. Central to this is the concept of *accurate dot product*. In particular, methods for solving and for verifying bounds on the solutions of linear systems of equations involve repeated computation of dot products. If outward roundings are done after each addition and multiplication in the dot product, then the widths of the intervals can rapidly increase during the computation, resulting in an enclosure for the dot product that is too wide to be useful. This is especially true if there is much cancelation during the computation, something that can happen for an ill-conditioned system, even if the components of the original vectors are points. Kulisch has advocated a *long accumulator*, described in [126] among other publications. The concept is that the sum in the dot product is stored in a register having so many digits that, when the final result is rounded to the usual number of digits, the result is the closest number less than or the closest number greater than the mathematically correct result. This long accumulator has been implemented in hardware in several machines and is implemented in software in C-XSC and the other XSC languages. However, the accurate dot product is relatively slow when implemented in software, and its necessity has been

[12] The SIAM 100-digit challenge was a series of 10 numerical problems posed by Nick Trefethen in which the goal was to supply 10 correct digits. A nominal prize was given for the best answers, and the contestants took it upon themselves to rigorously verify the correctness of the digits supplied.

controversial. Ogita et al. have proposed and analyzed methods for computing sums and dot products accurately and quickly using standard hardware, without a long accumulator [185]. Such algorithms do not always give a result that rounds to the nearest number to the correct result, but, with quadruple precision[13] give results that are within one or two units of the correct result except in certain cases; Rump et al. have a formalized analysis of this. This group has used these techniques in algorithms for verification of solutions of linear systems [186, 189, 226].

Accurate dot products in general will not help when there is interval uncertainty in the coefficients of the original linear system and the coefficients are assumed to vary independently.

Interval arithmetic is also available in the *Mathematica* [109] and Maple [37, 116] computer algebra systems. Finally, other and older implementations of interval arithmetic are described in [97, pp. 102–112].

[13] Now common and perhaps specified in the revision of the IEEE binary floating point standard.

Chapter 4

Further Properties of Interval Arithmetic

In Chapter 2 we introduced the definitions of the basic interval arithmetic operations. With proper understanding of the notation, it is possible to summarize them neatly in one line:

$$X \odot Y = \{x \odot y : x \in X,\ y \in Y\}, \quad \text{where } \odot \text{ can be } +,\ -,\ \cdot, \text{ or } /. \tag{4.1}$$

(Until Chapter 8, the quotient X/Y is defined only if $0 \notin Y$.) These definitions lead to a number of familiar looking algebraic properties. In addition, however, their set-theoretic character implies certain important relationships involving set containment. The material of the present chapter constitutes essential preparation for the study of interval functions in Chapter 5 and beyond.

4.1 Algebraic Properties

Commutativity and Associativity

It is easy to show that both interval addition and multiplication are commutative and associative; we have

$$X + Y = Y + X, \qquad X + (Y + Z) = (X + Y) + Z,$$
$$XY = YX, \qquad X(YZ) = (XY)Z$$

for any three intervals X, Y, and Z.

Additive and Multiplicative Identity Elements

The degenerate intervals 0 and 1 are additive and multiplicative identity elements in the system of intervals:

$$0 + X = X + 0 = X,$$
$$1 \cdot X = X \cdot 1 = X,$$
$$0 \cdot X = X \cdot 0 = 0$$

for any X.

Nonexistence of Inverse Elements

We caution that $-X$ is *not* an additive inverse for X in the system of intervals. Indeed, we have

$$X + (-X) = \left[\underline{X}, \overline{X}\right] + \left[-\overline{X}, -\underline{X}\right] = \left[\underline{X} - \overline{X}, \overline{X} - \underline{X}\right],$$

and this equals $[0, 0]$ *only* if $\underline{X} = \overline{X}$. If X does not have zero width, then

$$X - X = w(X)[-1, 1]. \tag{4.2}$$

Similarly, $X/X = 1$ only if $w(X) = 0$. In general,

$$X/X = \begin{cases} \left[\underline{X}/\overline{X}, \overline{X}/\underline{X}\right] & \text{if } 0 < \underline{X}, \\ \left[\overline{X}/\underline{X}, \underline{X}/\overline{X}\right] & \text{if } \overline{X} < 0. \end{cases} \tag{4.3}$$

We do not have additive or multiplicative inverses except for degenerate intervals. However, we always have the inclusions $0 \in X - X$ and $1 \in X/X$.

Exercise 4.1. Show that $X \cdot [-1, 1] = \max(|\underline{X}|, |\overline{X}|)[-1, 1]$. ☐

Exercise 4.2. Show that if $c > 0$, then $[c, c] \cdot \left[\underline{X}, \overline{X}\right] = \left[c\underline{X}, c\overline{X}\right]$ for any X. ☐

Subdistributivity

The distributive law

$$x(y + z) = xy + xz$$

of ordinary arithmetic also fails to hold for intervals. An easy counterexample can be obtained by taking $X = [1, 2]$, $Y = [1, 1]$, and $Z = -[1, 1]$:

$$\begin{aligned} X(Y + Z) &= [1, 2] \cdot ([1, 1] - [1, 1]) \\ &= [1, 2] \cdot [0, 0] \\ &= [0, 0], \end{aligned}$$

whereas (4.2) gives

$$\begin{aligned} XY + XZ &= [1, 2] \cdot [1, 1] - [1, 2] \cdot [1, 1] \\ &= [\min(1, 2), \max(1, 2)] - [\min(1, 2), \max(1, 2)] \\ &= [1, 2] - [1, 2] \\ &= [-1, 1]. \end{aligned}$$

However, there is a *subdistributive law*:

$$X(Y + Z) \subseteq XY + XZ. \tag{4.4}$$

We can see this in the example above. Full distributivity does hold in certain special cases. In particular, for any real number x we have

$$x(Y + Z) = xY + xZ. \tag{4.5}$$

Interval multiplication can be distributed over a sum of intervals as long as those intervals have the same sign:

$$X(Y + Z) = XY + XZ \quad \text{provided that} \quad YZ > 0. \tag{4.6}$$

Example 4.1. We have

$$[1, 2]\,([-3, -2] + [-5, -1]) = [-16, -3] = [1, 2][-3, -2] + [1, 2][-5, -1]$$

as an example of (4.6). □

Exercise 4.3. Prove properties (4.4)–(4.6). □

Cancellation

The *cancellation law*

$$X + Z = Y + Z \implies X = Y \tag{4.7}$$

holds for interval addition.

Exercise 4.4. Prove (4.7). □

Exercise 4.5. Show that multiplicative cancellation does *not* hold in interval arithmetic; that is, $ZX = ZY$ does *not* imply $X = Y$. □

We should emphasize that, with the identification of degenerate intervals and real numbers, interval arithmetic is an extension of real arithmetic. It reduces to ordinary real arithmetic for intervals of zero width.

4.2 Symmetric Intervals

An interval X is said to be *symmetric* if

$$\underline{X} = -\overline{X}. \tag{4.8}$$

Hence, $[-5, 5]$ and $[-\pi, \pi]$ are symmetric intervals. Any symmetric interval has midpoint 0. If X is symmetric, then according to (2.8) and (2.9),

$$|X| = \tfrac{1}{2}w(X) \quad \text{and} \quad X = |X|[-1, 1].$$

Exercise 4.6. Show that any interval X can be expressed as the sum of a real number (i.e., degenerate interval) and a symmetric interval:

$$X = m + W, \quad \text{where } m = m(X) \text{ and } W = \tfrac{1}{2}w(X)[-1, 1].$$ □

The rules of interval arithmetic are slightly simpler when symmetric intervals are involved. If X, Y, and Z are all symmetric, then

$$X + Y = X - Y = (|X| + |Y|)[-1, 1],$$
$$XY = |X||Y|[-1, 1],$$
$$X(Y \pm Z) = XY + XZ = |X|(|Y| + |Z|)[-1, 1].$$

If Y is symmetric and X is any interval, then

$$XY = |X|Y.$$

It follows that if Y and Z are symmetric, then

$$X(Y + Z) = XY + XZ$$

for any interval X. Compare this with the subdistributive property (4.4).

Numerical Exploitation of Interval Symmetry

Oliver Aberth's RANGE software, detailed in [1] and referenced several times later in this book, makes extensive use of the properties of symmetric intervals. See Appendix D.

In INTLAB, symmetric intervals can be exploited through the use of midpoint-radius arithmetic. Although there is, in general, overestimation in multiplication if midpoint-radius arithmetic is used, the operations are exact when the intervals are symmetric, and the resulting operations can be very fast. (See [223] for formulas for midpoint radius arithmetic, an analysis of its overestimation, and an analysis of speed that can be gained by using it.) Midpoint-radius arithmetic is used internally in certain places in INTLAB for matrix multiplication.[14] In INTLAB, midpoint-radius arithmetic can be selected to be used in those places by issuing the command `intvalinit('FastIVmult')`, while endpoint arithmetic[15] can be selected by issuing the INTLAB command `intvalinit('SharpIVmult')`. The type of multiplication in effect in those selected internal routines can be queried by issuing the MATLAB command `intvalinit('IVmult')`, which returns either `FastIVmult` or `SharpIVmult`.

4.3 Inclusion Isotonicity of Interval Arithmetic

Let $\odot$ stand for interval addition, subtraction, multiplication, or division. If A, B, C, and D are intervals such that

$$A \subseteq C \quad \text{and} \quad B \subseteq D,$$

then

$$A \odot B \subseteq C \odot D.$$

These relations follow directly from the definitions given in Chapter 2. However, these relations are not purely algebraic; they serve to connect the algebraic and set properties of

[14] See our Chapter 7 for more about linear algebra with intervals.

[15] Executed with our formulas (2.21), (2.22), (2.23) or Table 2.1, and (2.24) on p. 13.

interval arithmetic. Interval arithmetic is said to be *inclusion isotonic*. We will see important applications of this idea later on.

In the next chapter, we extend the concept of interval expressions to include functions, such as $\sin x$ and e^x.

Chapter 5

Introduction to Interval Functions

In Example 3.1, we used the simple formula

$$A = L \cdot W$$

to bound the area of a rectangle in terms of given bounds on its side lengths. Such a formula can be used to define an *interval-valued function* A of two interval variables L and W.

This chapter treats the basics of interval-valued functions. Standard functions such as exponentials can be applied directly to interval arguments, with interval results. More importantly, we will need the general notion of an interval-valued mapping to progress beyond interval arithmetic and into the realm of interval analysis. The reader can find a brief review of function terminology in Appendix A.

5.1 Set Images and United Extension

Let f be a real-valued function of a single real variable x. Ultimately, we would like to know the precise range of values taken by $f(x)$ as x varies through a given interval X. In other words, we would like to be able to find the image of the set X under the mapping f:

$$f(X) = \{f(x) : x \in X\}. \tag{5.1}$$

More generally, given a function $f = f(x_1, \dots, x_n)$ of several variables, we will wish to find the image set

$$f(X_1, \dots, X_n) = \{f(x_1, \dots, x_n) : x_1 \in X_1, \dots, x_n \in X_n\}, \tag{5.2}$$

where $X_1, \dots, X_n$ are specified intervals. Much of the present chapter will be devoted to this goal.

The set images described above can be characterized in a slightly different manner. We state the next definition mainly for completeness and for the convenient terminology that it provides.

Definition 5.1. Let $g\colon M_1 \to M_2$ be a mapping between sets M_1 and M_2, and denote by $S(M_1)$ and $S(M_2)$ the families of subsets of M_1 and M_2, respectively. The *united extension* of g is the set-valued mapping $\bar{g}\colon S(M_1) \to S(M_2)$ such that

$$\bar{g}(X) = \{g(x)\colon x \in X,\ X \in S(M_1)\}. \tag{5.3}$$

The mapping $\bar{g}$ is sometimes of interest as a single-valued mapping on $S(M_1)$ with values in $S(M_2)$. For our purposes, however, it is merely necessary to note that[16]

$$\bar{g}(X) = \bigcup_{x \in X}\{g(x)\},$$

i.e., that $\bar{g}(X)$ contains precisely the same elements as the set image $g(X)$. For this reason, and because the usage is common, we shall apply the term *united extension* to set images such as those described in (5.1) and (5.2).

5.2 Elementary Functions of Interval Arguments

For some functions, (5.1) is rather easy to compute. For example, consider

$$f(x) = x^2, \qquad x \in \mathbb{R}.$$

If $X = \left[\underline{X}, \overline{X}\right]$, it is evident that the set

$$f(X) = \{x^2\colon x \in X\} \tag{5.4}$$

can be expressed as

$$f(X) = \begin{cases} \left[\underline{X}^2, \overline{X}^2\right], & 0 \le \underline{X} \le \overline{X}, \\ \left[\overline{X}^2, \underline{X}^2\right], & \underline{X} \le \overline{X} \le 0, \\ \left[0, \max\left\{\underline{X}^2, \overline{X}^2\right\}\right], & \underline{X} < 0 < \overline{X}. \end{cases} \tag{5.5}$$

We will use (5.4) as the definition of X^2. This is *not* the same as $X \cdot X$. For instance,

$$[-1, 1]^2 = [0, 1], \qquad \text{whereas} \qquad [-1, 1] \cdot [-1, 1] = [-1, 1].$$

However, $[-1, 1]$ *does contain* $[0, 1]$. The overestimation when we compute a bound on the range of X^2 as $X \cdot X$ is due to the phenomenon of *interval dependency*. Namely, if we assume x is an unknown number known to lie in the interval X, then, when we form the product $x \cdot x$, the x in the second factor, although known only to lie in X must be the same as the x in the first factor, whereas, in the definition of the interval product $X \cdot X$, it is assumed that the values in the first factor and the values in the second factor vary independently.

Interval dependency is a crucial consideration when using interval computations. It is a major reason why simply replacing floating point computations by intervals in an existing algorithm is not likely to lead to satisfactory results.

[16]This accounts for the term *united*. The united extension $\bar{g}$ is an extension of g in the sense that we can identify singleton sets $\{x\} \in S(M_1)$ with their corresponding elements $x \in M_1$.

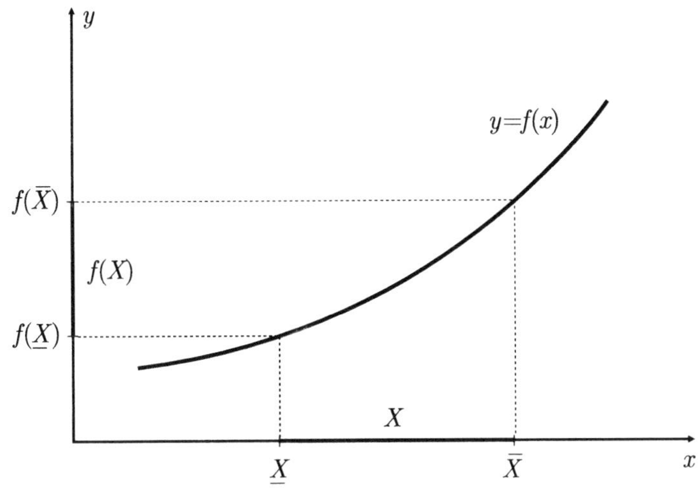

Figure 5.1. *The image of an interval X under an increasing function $f(x)$.*

Exercise 5.1. Within a MATLAB command window and with INTLAB running, issue the sequence of commands:

```
intvalinit('DisplayInfSup')
x = infsup(-1,1)
x^2
x*x
```

What does INTLAB return? Does it account for interval dependency in computing bounds on the range of $f(x) = x^2$? □

Use of Monotonic Functions

Let us apply some other familiar functions to interval arguments. The reasoning is particularly straightforward with functions $f(x)$ that happen to be *monotonic*, i.e., either increasing or decreasing with increasing x. An increasing function is depicted in Figure 5.1. Note that it maps an interval $X = [\underline{X}, \overline{X}]$ into the interval

$$f(X) = [f(\underline{X}), f(\overline{X})].$$

Let us take an example. As x varies through an interval $X = [\underline{X}, \overline{X}]$, the exponential function

$$f(x) = \exp(x) = e^x \qquad (x \in \mathbb{R})$$

takes values from $\exp(\underline{X})$ to $\exp(\overline{X})$. That is, we can define

$$\exp(X) = [\exp(\underline{X}), \exp(\overline{X})]. \tag{5.6}$$

The situation is similar for the logarithmic function

$$f(x) = \log x \qquad (x > 0);$$

we have

$$\log X = \left[\log \underline{X}\,,\ \log \overline{X}\right] \quad \text{for} \quad \underline{X} > 0. \tag{5.7}$$

Exercise 5.2. The capacity of a binary communication channel is given by the Hartley–Shannon equation $C = B \log_2(1 + S/N)$ (bits/s), where B is the channel bandwidth, and S/N is the signal-to-noise ratio. The formula $(S/N)_{\text{dB}} = 10 \log_{10}(S/N)$ defines the signal-to-noise ratio in decibels (dB). Find the possible range of C for a set of channels that may range in bandwidth from 4–8 kHz and in S/N from 20–25 dB. □

We could give many examples in this same vein. The square root of an interval is given by

$$\sqrt{X} = \left[\sqrt{\underline{X}}\,,\ \sqrt{\overline{X}}\right] \quad \text{for} \quad \underline{X} \geq 0. \tag{5.8}$$

Exercise 5.3. Find the diagonal of the rectangle discussed in Example 3.1. □

Exercise 5.4. For an antenna located h feet above the ground, the distance in miles to the line-of-sight radio horizon is given by $d = \sqrt{2h}$. Hence, if a transmitting antenna sits atop a tower of height h_1 ft, and a receiving antenna sits atop a building at height h_2 ft, their maximum allowed separation for line-of-sight communication is $s = \sqrt{2h_1} + \sqrt{2h_2}$ miles. Evaluate the possible range of s values for towers ranging in height from 1000 to 1500 ft and buildings ranging in height from 20 to 50 ft. □

The more general exponential function $f(x) = x^y$ with $x > 0$ and $y > 0$ leads us to write

$$X^y = \left[\underline{X}^y\,,\ \overline{X}^y\right] \quad \text{for} \quad \underline{X} > 0 \text{ and } y > 0. \tag{5.9}$$

All these functions have been increasing. With decreasing functions, we must be careful to order the endpoints correctly. For example, as x increases from $\underline{X}$ to $\overline{X}$, the values of $\exp(-x)$ decrease from $\exp(-\underline{X})$ to $\exp(-\overline{X})$. Therefore,

$$\exp(-X) = \left[\exp(-\overline{X})\,,\ \exp(-\underline{X})\right]. \tag{5.10}$$

We should also note that some restrictions of a nonmonotonic function could be monotonic. The function f given by

$$f(x) = \sin x, \qquad x \in \mathbb{R},$$

is not monotonic, but its restriction f_A to the set $A = [-\frac{\pi}{2}, \frac{\pi}{2}]$ is increasing. Hence,

$$\sin X = \left[\sin \underline{X}\,,\ \sin \overline{X}\right] \quad \text{for } X \subseteq [-\tfrac{\pi}{2}, \tfrac{\pi}{2}]. \tag{5.11}$$

We can make use of known properties of the sine function to find interval extensions for any interval argument. We know, for instance, that $\sin(k\pi + x) = (-1)^k \sin x$ and $\sin x$ is in $[-1, 1]$ for all real x.

INTLAB Applications

INTLAB's computation of tight bounds on the ranges of elementary functions (sin, exp, etc.) is described in [225]. Within GlobSol, a system for global optimization we will reference later, the implementation of computation of bounds on the ranges of the elementary functions is explained in [82, 101].

Exercise 5.5. A prism has angle α. The index of refraction n of the glass can be determined experimentally by measuring the minimum deviation angle δ and using the formula

$$n = \frac{\sin \frac{1}{2}(\delta + \alpha)}{\sin \frac{1}{2}\alpha}.$$

If measurements indicate that $\alpha = 60° \pm 0.5°$ and $\delta = 45° \pm 0.5°$, find an interval enclosing n. □

Exercises 5.3, 5.4, and 5.5 can be done easily in INTLAB, since INTLAB has preprogrammed bounds on the ranges for many commonly used functions. See Appendix E.

The way that INTLAB computes bounds on the range of a function can be examined by looking at the corresponding m-file. For instance, the INTLAB program that computes the range of $\sqrt{x}$ over an interval X is found in

```
<intlabdir>/intval/@intval/sqrt.m
```

where `<intlabdir>` is the directory (or folder) in which INTLAB is installed. This m-file is somewhat more complicated than the explanation here might suggest, because it handles cases where the output is complex,[17] it handles cases where the argument passed to `sqrt` is an array or a sparse matrix, and it catches cases where difficulties are encountered. For example, in using `sqrt.m`, we see the following in a MATLAB command window:

```
>> x = infsup(-1,-1)
intval x = [ -1.0000 , -1.0000 ]
>> sqrt(x)
Warning: SQRT: Real interval input out of range changed to be
complex
> In intval.sqrt at 140
intval ans = [ -0.0001 + 0.9999i , 0.0001 + 1.0001i ]
```

However, for nonnegative purely real arguments, the INTLAB routine `sqrt.m` calls a routine `sqrt_rnd.m`, found in

```
<intlabdir>/intval/@intval/private/sqrt_rnd.m
```

twice, once for the lower endpoint and once for the upper endpoint; the function `sqrt_rnd.m` simply evaluates the square root using MATLAB's `sqrt` function for noninterval arguments then adjusts the value downward or upward to make sure the machine-representable number returned is a lower bound or upper bound (as requested) on the actual value.

Exercise 5.6. For the function $f(x) = x^2$, INTLAB uses the program in the m-file

```
<intlabdir>/intval/@intval/sqr.m
```

[17] In fact, consistent with the standard functions in MATLAB, INTLAB implements complex extensions of the standard functions. An explanation is given in [29].

Print out this file and annotate it, identifying where the lower and upper bounds on X^2 are computed. □

Looking back, we see that equations (5.6)–(5.11) define interval-valued functions of interval variables. We stress that evaluation of one of these functions F, as indicated, at a given interval X, yields precisely the set $f(X)$ described by equation (5.1). The serious issue of interval dependency is lurking, however, as we are about to see.

5.3 Interval-Valued Extensions of Real Functions

In the last section, we were able to define a few interval-valued functions. We did this by selecting a real-valued function f and computing the range of values $f(x)$ taken as x varied through some interval X. By definition, the result was equal to the set image $f(X)$.

Here we will consider a different process: that of *extending* a given real-valued function f by applying its *formula* directly to interval arguments.

Formulas and Interval Extensions

Let us begin with an example. Consider the real-valued function f given by

$$f(x) = 1 - x, \qquad x \in \mathbb{R}. \tag{5.12}$$

Note carefully that a function is defined by two things: (1) a domain over which it acts, and (2) a rule that specifies how elements of that domain are mapped under the function. Both of these are specified in (5.12): the elements of Dom f are real numbers x, and the *mapping rule* is $x \mapsto 1 - x$. Taken in isolation, the entity

$$f(x) = 1 - x \tag{5.13}$$

is a *formula*—not a function. Often this distinction is ignored; in many elementary math books, for example, we would interpret (5.13) as a function whose domain should be taken as the largest possible set over which the formula makes sense (in this case, all of $\mathbb{R}$). However, we will understand that Dom f is just as essential to the definition of f as is the formula $f(x)$.

Now suppose we take the formula (5.13) that describes function (5.12) and apply it to interval arguments. The resulting interval-valued function

$$F(X) = 1 - X, \qquad X = [\underline{X}, \overline{X}], \tag{5.14}$$

is *an extension* of the function (5.12): we have enlarged the domain to include nondegenerate intervals X as well as the degenerate intervals $x = [x, x]$.

Definition 5.2. We say that F is an *interval extension* of f, if for degenerate interval arguments, F agrees with f:

$$F([x, x]) = f(x). \tag{5.15}$$

Let us compare $F(X)$ with the set image $f(X)$. We have, according to the laws of interval arithmetic,

$$F(X) = [1, 1] - [\underline{X}, \overline{X}] = [1, 1] + [-\overline{X}, -\underline{X}] = [1 - \overline{X}, 1 - \underline{X}].$$

On the other hand, as x increases through the interval $[\underline{X}, \overline{X}]$, the values $f(x)$ given by (5.12) clearly decrease from $1 - \underline{X}$ to $1 - \overline{X}$; by definition then,

$$f(X) = \left[1 - \overline{X}, 1 - \underline{X}\right].$$

In this example, we have $F(X) = f(X)$: this particular extension of f, obtained by applying formula (5.13) directly to interval arguments, yields the desired set image (5.1). In other words, we have found the united extension of f:

$$f(X) = 1 - X.$$

Unfortunately, as we are about to see, the situation is not so simple in general.

A More Interesting Example

Consider next the real-valued function

$$f(x) = x(1 - x), \qquad x \in [0, 1]. \tag{5.16}$$

For reasons that will become apparent, we use this to write *two* real-valued functions:

$$f(x) = x(1 - x), \qquad x \in [0, 1], \tag{5.17}$$

and

$$g(x) = x - x^2, \qquad x \in [0, 1]. \tag{5.18}$$

These are mathematically equal, because *in ordinary real arithmetic* we have

$$x(1 - x) = x - x^2.$$

As x increases from 0 to 1, the values $f(x)$ and $g(x)$ both increase from 0 to $\frac{1}{4}$ then decrease back to 0. Therefore,

$$f([0, 1]) = g([0, 1]) = \left[0, \tfrac{1}{4}\right]. \tag{5.19}$$

Let us form interval-valued extensions of f and g:

$$F(X) = X \cdot (1 - X), \qquad X = \left[\underline{X}, \overline{X}\right], \tag{5.20}$$

and

$$G(X) = X - X^2, \qquad X = \left[\underline{X}, \overline{X}\right]. \tag{5.21}$$

Remembering that $X^2 \neq X \cdot X$ in interval arithmetic, we work out the details of each extension separately for an interval $X = [\underline{X}, \overline{X}] \subseteq [0, 1]$:

$$\begin{aligned}
F(X) &= \left[\underline{X}, \overline{X}\right] \cdot \left([1, 1] - [\underline{X}, \overline{X}]\right) \\
&= \left[\underline{X}, \overline{X}\right] \cdot \left([1, 1] + \left[-\overline{X}, -\underline{X}\right]\right) \\
&= \left[\underline{X}, \overline{X}\right] \cdot \left(\left[1 - \overline{X}, 1 - \underline{X}\right]\right) \\
&= [\min S, \max S],
\end{aligned}$$

where

$$S = \{\underline{X}(1-\overline{X}),\ \underline{X}(1-\underline{X}),\ \overline{X}(1-\overline{X}),\ \overline{X}(1-\underline{X})\},$$

while

$$\begin{aligned} G(X) &= [\underline{X},\overline{X}] - [\underline{X},\overline{X}]^2 \\ &= [\underline{X},\overline{X}] - \left[\underline{X}^2,\overline{X}^2\right] \\ &= [\underline{X},\overline{X}] + \left[-\overline{X}^2,-\underline{X}^2\right] \\ &= \left[\underline{X}-\overline{X}^2,\overline{X}-\underline{X}^2\right]. \end{aligned}$$

Putting, say, $X = [0, 1]$, we see that $G(X) \neq F(X)$; the two formulas involved in (5.17) and (5.18)—again, *identical* in ordinary arithmetic—give rise to *different* extensions, and neither of these maps the interval $[0, 1]$ into the interval $\left[0, \frac{1}{4}\right]$. We have $F([0, 1]) = [0, 1]$ and $G([0, 1]) = [-1, 1]$.

We stress that two expressions can be equivalent in real arithmetic but not equivalent in interval arithmetic. This is due to the lack of distributivity and additive and multiplicative inverses in interval arithmetic.

It turns out that the united extension of the original function f arises from use of a third equivalent formula:

$$h(x) = \tfrac{1}{4} - \left(x - \tfrac{1}{2}\right)^2.$$

We get

$$\begin{aligned} H(X) &= \tfrac{1}{4} - \left(X - \tfrac{1}{2}\right)^2 \\ &= \left[\tfrac{1}{4},\tfrac{1}{4}\right] - \left([\underline{X},\overline{X}] - \left[\tfrac{1}{2},\tfrac{1}{2}\right]\right)^2 \\ &= \left[\tfrac{1}{4},\tfrac{1}{4}\right] - \left[\underline{X}-\tfrac{1}{2},\overline{X}-\tfrac{1}{2}\right]^2 \\ &= \left[\tfrac{1}{4},\tfrac{1}{4}\right] - \begin{cases} \left[(\underline{X}-\frac{1}{2})^2,(\overline{X}-\frac{1}{2})^2\right], & \underline{X} \geq \frac{1}{2}, \\ \left[(\overline{X}-\frac{1}{2})^2,(\underline{X}-\frac{1}{2})^2\right], & \overline{X} \leq \frac{1}{2}, \\ \left[0,\max\{(\underline{X}-\frac{1}{2})^2,(\overline{X}-\frac{1}{2})^2\}\right], & \underline{X} < \frac{1}{2} < \overline{X}, \end{cases} \end{aligned}$$

so that

$$H(X) = \begin{cases} \left[\frac{1}{4}-(\overline{X}-\frac{1}{2})^2,\frac{1}{4}-(\underline{X}-\frac{1}{2})^2\right], & \underline{X} \geq \frac{1}{2}, \\ \left[\frac{1}{4}-(\underline{X}-\frac{1}{2})^2,\frac{1}{4}-(\overline{X}-\frac{1}{2})^2\right], & \overline{X} \leq \frac{1}{2}, \\ \left[\frac{1}{4}-\max\{(\underline{X}-\frac{1}{2})^2,(\overline{X}-\frac{1}{2})^2\},\frac{1}{4}\right], & \underline{X} < \frac{1}{2} < \overline{X}. \end{cases}$$

It is easily verified that this is $f(X)$. We have $H([0, 1]) = \left[0, \frac{1}{4}\right]$.

Exercise 5.7. Let $A_1 = [0, 1]$, $A_2 = [1, 2]$, and $A_3 = [-2, 1]$. Evaluate the two expressions

$$(A_1 - A_2) \cdot (A_3/A_2) \quad \text{and} \quad (A_1/A_2 - 1) \cdot A_3$$

and compare the results. Note that these expressions would be equivalent in ordinary arithmetic. □

Exercise 5.8. Show that there is never a *unique* interval extension of a given real function. □

Exercise 5.9. Evaluation of $(x-1)^6$ for values near $x = 1$ is sometimes used in traditional numerical analysis courses to illustrate the effects of roundoff error. In particular, if we define

$$\begin{aligned} f(x) &= x^6 - 6x^5 + 15x^4 - 20x^3 + 15x^2 - 6x + 1, \\ g(x) &= 1 + x(-6 + x(15 + x(-20 + x(15 + x(-6 + x))))), \\ h(x) &= (x-1)^6, \end{aligned}$$

then f, g, and h define the same real-valued function. However, if one evaluates f, g, and h using floating point arithmetic at points near $x = 1$ and plots the results, one obtains very different graphs. A related phenomenon occurs with interval values.

1. Use floating point evaluation within MATLAB to make approximate plots of f, g, and h at 100 points in $[0.999, 1.001]$.
2. Using INTLAB, set `X = infsup(0.999,1.001)` and evaluate f, g, h over X.
3. Compare parts 1 and 2. Observe the effects of roundoff error in the floating point plots of f, g, and h. Which expression is most prone to roundoff "noise"? Does the susceptibility to roundoff noise correlate with the width of the corresponding interval extension? □

5.4 The Fundamental Theorem and Its Applications

Subset Property of United Extension

Looking back at (5.3), we see that the united extension $\bar{g}$ has the following *subset property*:

$$X, Y \in S(M_1) \text{ with } X \subseteq Y \implies \bar{g}(X) \subseteq \bar{g}(Y). \tag{5.22}$$

We will return to this observation momentarily.

Interval Extensions of Multivariable Functions

So far, we have limited ourselves to functions of a single interval variable X. In principle, there is no reason we should avoid more general functions,

$$f = f(X_1, \ldots, X_n),$$

depending on n interval variables $X_1, \ldots, X_n$.

Definition 5.3. By an *interval extension* of f, we mean an interval-valued function F of n interval variables $X_1, \ldots, X_n$ such that for real arguments $x_1, \ldots, x_n$ we have

$$F(x_1, \ldots, x_n) = f(x_1, \ldots, x_n). \tag{5.23}$$

That is, if the arguments of F are degenerate intervals, then the left-hand side of (5.23) is a degenerate interval equal to the right-hand side.

As simple yet crucial examples of two-variable functions, we have the operations of interval arithmetic. Let us take interval addition as an example. Although we do not normally do so, we could denote this binary operation as a function of two variables:

$$f(X_1, X_2) = X_1 + X_2 = \{x_1 + x_2 \colon x_1 \in X_1,\ x_2 \in X_2\}.$$

The notation is consistent with our previous usage: indeed, it follows from the set definition on the right that interval addition is the united extension of the function

$$f(x_1, x_2) = x_1 + x_2$$

that describes ordinary numerical addition. The operations of interval arithmetic are all defined in this way—recall (4.1). The essential observation is that the interval arithmetic functions are united extensions of the corresponding real arithmetic functions.

Inclusion Isotonicity

Definition 5.4. We say that $F = F(X_1, \ldots, X_n)$ is *inclusion isotonic* if

$$Y_i \subseteq X_i \text{ for } i = 1, \ldots, n \implies F(Y_1, \ldots, Y_n) \subseteq F(X_1, \ldots, X_n).$$

Observe carefully that united extensions, which all have the subset property, are inclusion isotonic. In particular, then, the operations of interval arithmetic must satisfy

$$Y_1 \subseteq X_1,\ Y_2 \subseteq X_2 \implies Y_1 \odot Y_2 \subseteq X_1 \odot X_2. \tag{5.24}$$

We made this observation in section 4.3. Here, we will go a step further.

Rational Interval Functions

Definition 5.5. A *rational interval function* is an interval-valued function whose values are defined by a specific finite sequence of interval arithmetic operations.

Example 5.1. Consider the function given by

$$F(X_1, X_2) = ([1, 2]X_1 + [0, 1])\, X_2.$$

The computation of $F(X_1, X_2)$ can be broken down into the finite sequence of interval arithmetic operations described by

$$\begin{aligned} T_1 &= [1, 2]X_1 && \text{(one interval multiplication)},\\ T_2 &= T_1 + [0, 1] && \text{(one interval addition)},\\ F(X_1, X_2) &= T_2 X_2 && \text{(one interval multiplication)}. \end{aligned}$$

Hence F is a rational interval function. □

By nature, a rational interval function must arise as an extension of some real-valued function. In the example above, $F(X_1, X_2)$ extends a real function of the form

$$f(x_1, x_2) = (c_1 x_1 + c_2)x_2,$$

where c_1 and c_2 are constants. The function

$$F(X) = m(X) + \tfrac{1}{2}(X - m(X))$$

is *not* a rational interval function (even though its computation entails only a finite number of operations) because $m(X) = (\underline{X} + \overline{X})/2$ is not an interval arithmetic operation.

Lemma 5.1. All rational interval functions are inclusion isotonic.

Proof. The lemma follows by finite induction from (5.24) and the transitivity of the relation $\subseteq$. $\square$

The Fundamental Theorem

We are now ready to state a central theorem.

Theorem 5.1 (Fundamental Theorem of Interval Analysis [148]). If F is an inclusion isotonic interval extension of f, then

$$f(X_1, \ldots, X_n) \subseteq F(X_1, \ldots, X_n).$$

Proof. By definition of an interval extension, $f(x_1, \ldots, x_n) = F(x_1, \ldots, x_n)$. If F is inclusion isotonic, then the value of f is contained in the interval $F(X_1, \ldots, X_n)$ for every $(x_1, \ldots, x_n)$ in $(X_1, \ldots, X_n)$. $\square$

Natural Interval Extensions

In section 5.3, we obtained extensions F of real rational functions f by replacing (1) the real variable x with an interval variable X and (2) the real arithmetic operations with corresponding interval operations. The result F is called a *natural interval extension* of f. The same procedure can be performed with functions of n variables, and our general observation from section 5.3 bears repeating:

> *Two rational expressions which are equivalent in real arithmetic may not be equivalent in interval arithmetic.*

However, since rational interval functions are inclusion isotonic, we have the following corollary to Theorem 5.1.

Corollary 5.1. If F is a rational interval function and an interval extension of f, then

$$f(X_1, \ldots, X_n) \subseteq F(X_1, \ldots, X_n).$$

That is, an interval value of F contains the range of values of the corresponding real function f when the real arguments of f lie in the intervals shown.

Therefore, we have a means for the finite evaluation of upper and lower bounds on the ranges of values of real rational functions.

Example 5.2. Consider the polynomial

$$p(x) = 1 - 5x + \tfrac{1}{3}x^3. \tag{5.25}$$

Suppose we wish to know the range of values of $p(x)$ when $2 \le x \le 3$. Since the interval polynomial

$$P(X) = 1 - 5X + \tfrac{1}{3}X \cdot X \cdot X \tag{5.26}$$

is a natural interval extension of $p(x)$, the actual range of $p(x)$ values is contained in the interval

$$P([2,3]) = 1 - 5[2,3] + \tfrac{1}{3}[8,27] = \left[-\tfrac{34}{3}, 0\right].$$

A different extension can be obtained by first rewriting $p(x)$ as

$$q(x) = 1 - x(5 - x^2/3).$$

A natural interval extension of this is

$$Q(X) = 1 - X(5 - X \cdot X/3), \tag{5.27}$$

and a straightforward calculation yields

$$Q([2,3]) = [-10, -3].$$

This narrower interval provides a better estimate for the desired range of values of $p(x)$. □

Using ordinary calculus, it is easy to verify that the exact range of values for $p(x)$ in the preceding example is

$$p(x) \in \left[-\tfrac{10}{3}\sqrt{5} + 1\,,\ -5\right] = [-6.453559\ldots, -5] \quad \text{for} \quad x \in [2,3]. \tag{5.28}$$

Exercise 5.10. Verify (5.28). □

However, an important point is that we can find intervals containing the exact range of values without calculus — just by a single evaluation of an interval expression.

We cannot find (5.28) exactly by *any* finite sequence of arithmetic operations with rational numbers, since $\sqrt{5}$ is irrational. In Chapter 6, we will discuss interval methods for computing convergent sequences of upper and lower bounds to exact ranges of values.

At this time, we point out that for polynomials, the nested form

$$A_0 + X(A_1 + X(A_2 + \cdots + X(A_n)\cdots)$$

usually gives better (and never gives worse) results than the sum-of-powers form

$$A_0 + A_1X + A_2X \cdot X + \cdots + A_nX \cdot X \cdot \cdots \cdot X.$$

This statement holds by (4.4).

It turns out that any natural interval extension of a rational function *in which each variable occurs only once* (if at all) *and to the first power only* will compute the *exact range* of values provided that no division by an interval containing zero occurs. Equation (5.14) was a simple example of such an extension. A proof will be provided in Chapter 6.

Exercise 5.11. Show by a counterexample that the interval function

$$F(X) = m(X) + \tfrac{1}{2}(X - m(X))$$

is not inclusion isotonic. □

Exercise 5.12. The resistance of a parallel combination of two electrical resistors is given by the formula $R^{-1} = R_1^{-1} + R_2^{-1}$. If R_1 and R_2 have nominal values of 47 Ω and 10 Ω, respectively, and manufacturing tolerances of 10%, find an interval enclosure for R. Repeat using the formula $R = (R_1 R_2)/(R_1 + R_2)$, which is equivalent in real arithmetic. □

Exercise 5.13. Find $f([0, 3])$ if $f(x) = \frac{1}{3}x^3 - 2x$. Compare with enclosures found by evaluating some interval extensions of f. □

A Further Property of Natural Extensions

Suppose that $F(X)$ is a natural interval extension of a real rational function $f(x)$, and suppose we can evaluate $F(X_0)$ for some interval X_0 without encountering a division by an interval containing zero. Then the range of values of $f(x)$ for x in X_0 is bounded by

$$\underline{F(X_0)} \le f(x) \le \overline{F(X_0)}.$$

Hence, $f(x)$ cannot have any poles in X_0. Furthermore, by the inclusion isotonicity of the interval arithmetic operations (and hence of $F(X)$), no division by an interval containing zero will occur during the evaluation of the same expression for $F(X)$ for any $X \subseteq X_0$. In other words, if $F(X_0)$ is defined, then $F(X)$ also is for every $X \subseteq X_0$.

5.5 Remarks on Numerical Computation

As we close this chapter, let us remember that interval methods are for use with computers. In particular, IA is an integral part of these methods. A principal goal of interval computation is to use computers to produce pairs of numbers, between which there are *exact* solutions of various kinds of computational problems.

In section 5.2 we defined X^2, where X is an interval. Higher integral powers X^n are defined by

$$X^n = \begin{cases} \left[\underline{X}^n, \overline{X}^n\right] & \text{if } \underline{X} > 0 \text{ or } n \text{ is odd}, \\ \left[\overline{X}^n, \underline{X}^n\right] & \text{if } \overline{X} < 0 \text{ and } n \text{ is even}, \\ [0, |X|^n] & \text{if } 0 \in X \text{ and } n \text{ is even}. \end{cases} \tag{5.29}$$

This way we have

$$X^n = \{x^n : x \in X\}$$

for each n. To take the limited precision of machine arithmetic into account, we can treat $\underline{X}$, $\overline{X}$, and $|X|$ as degenerate intervals and compute $\underline{X}^n$, $\overline{X}^n$, and $|X|$ in IA, using the appropriate endpoint obtained in this way in (5.29). Thus we can compute, in each case, a machine interval containing the exact result.

Exercise 5.14. Study the INTLAB file

```
<intlabdir>/intval/@intval/power.m
```

Explain where (5.29) is implemented. What other necessary complication exists in this function? □

In section 5.2 we used monotonicity to calculate the images of intervals X under certain common functions. We saw examples of the fact that if f is increasing, then

$$f(X) = \left[f(\underline{X}), f(\overline{X}) \right],$$

while if f is decreasing, then

$$f(X) = \left[f(\overline{X}), f(\underline{X}) \right].$$

These united extension values are, of course, inaccessible in practice because the required endpoints cannot be computed exactly. However, we can extend the machine-computed endpoints outward in each case by an amount just great enough to include the approximation error for each such function. This is done automatically by interval software packages such as INTLAB. In this way we can compute, on the machine, intervals containing the exact ranges of values. The resulting functions are inclusion isotonic.

Using similar approaches, we can construct inclusion isotonic interval extensions of all functions commonly used in computing.

In the next chapter, we introduce theory and practice for computing refinements of interval extensions, such that we enclose the corresponding united extensions arbitrarily closely.

Chapter 6

Sequences of Intervals and Interval Functions

6.1 A Metric for the Set of Intervals

Review of Convergence and Continuity

In real analysis, we call a sequence $\{x_k\}$ *convergent* if there exists a real number x^* such that for every $\varepsilon > 0$ there is a natural number $N = N(\varepsilon)$ such that

$$|x_k - x^*| < \varepsilon \tag{6.1}$$

whenever $k > N$. In this case, we write

$$x^* = \lim_{k \to \infty} x_k \quad \text{or} \quad x_k \to x^*$$

and refer to x^* as the *limit* of $\{x_k\}$. The quantity $|x_k - x^*|$ in (6.1) is just the distance between x_k and x^* as measured along the real line.

The notion of sequence convergence appears throughout mathematics. One can consider convergent sequences of complex numbers, real-valued functions of a real variable, etc. All that is needed is a suitable measure of "distance" between the objects of interest.

Let us also recall the definition of ordinary function continuity. We say that $f(x)$ is *continuous* at a point x_0 if for every $\varepsilon > 0$ there is a positive number $\delta = \delta(\varepsilon)$ such that

$$|f(x) - f(x_0)| < \varepsilon \tag{6.2}$$

whenever $|x - x_0| < \delta$. In this case, $|f(x) - f(x_0)|$ is the distance between the y-axis points $f(x)$ and $f(x_0)$, corresponding to the distance $|x - x_0|$ along the x-axis.

It is said that convergence and continuity are the two central concepts of analysis. We see, in turn, how they both hinge on having a suitable way to express distance. The need to discuss convergence and continuity outside of ordinary real analysis has led to a powerful generalization of the distance idea.

Notion of Metric

Let S be any set, and suppose that a real-valued function d is defined such that for any two elements $x, y \in S$ the following statements hold:

1. $d(x, y) = 0$ if and only if $x = y$;

2. $d(x, y) = d(y, x)$;

3. $d(x, y) \le d(x, z) + d(z, y)$ for any $z \in S$.

These can be regarded as the essential characteristics of distance between the objects x and y, and they certainly hold in the real number system when $x, y \in \mathbb{R}$ and $d(x, y) = |x - y|$. The function d is called a *metric on* S, and S is known as a *metric space*.

It is not our purpose to develop the theory of metric spaces. However, we must discuss continuity and convergence in the context of interval analysis, and for this we require a suitable metric. We will use

$$d(X, Y) = \max\{|\underline{X} - \underline{Y}|, |\overline{X} - \overline{Y}|\} \tag{6.3}$$

as a measure of distance between two intervals $X = [\underline{X}, \overline{X}]$ and $Y = [\underline{Y}, \overline{Y}]$.

Example 6.1. The distance between the intervals $X = [1, 2]$ and $X = [3, 5]$ is $d(X, Y) = \max\{|1 - 3|, |2 - 5|\} = \max\{2, 3\} = 3$. □

Convergence and Continuity in Interval Mathematics

Let $\{X_k\}$ be a sequence of intervals. We say that $\{X_k\}$ is convergent if there exists an interval X^* such that for every $\varepsilon > 0$, there is a natural number $N = N(\varepsilon)$ such that $d(X_k, X^*) < \varepsilon$ whenever $k > N$. Here d is given by (6.3). As in the case of real sequences, we write

$$X^* = \lim_{k \to \infty} X_k \quad \text{or} \quad X_k \to X^*$$

and refer to X^* as the limit of $\{X_k\}$.

Exercise 6.1. Show that $X_k \to X$ if and only if $\underline{X}_k \to \underline{X}$ and $\overline{X}_k \to \overline{X}$ in the sense of real sequences. Also show that the limit of a convergent sequence of intervals is unique. □

Exercise 6.2. (a) Let F be an interval-valued function of the interval variable X. What does it mean for F to be continuous at some X_0? (b) Propose a definition of *uniform* continuity for an interval function. □

It is not hard to show that the interval arithmetic operations and rational interval functions are continuous if no division by an interval containing zero occurs.

Exercise 6.3. Show that interval addition is continuous. □

Isometric Embedding

We know that the interval number system represents an extension of the real number system. In fact, the correspondence $[x, x] \leftrightarrow x$ of (2.14) can be regarded as a function or *mapping* between the two systems. This mapping preserves distances between corresponding objects: we have

$$d([x, x], [y, y]) = \max\{|x - y|, |x - y|\} = |x - y|$$

for any x and y. For this reason, it is called an *isometry*, and we say that the real line is "isometrically embedded" in the metric space of intervals.

Exercise 6.4. Show that (6.3) has the following additional properties: (a) $d(X+Z, Y+Z) = d(X, Y)$; (b) $d(X, Y) \le w(Y)$ when $X \subseteq Y$; (c) $d(X, 0) = |X|$. □

Exercise 6.5. In INTLAB, (6.3) is implemented with the function `qdist(X,Y)`:

```
>> X = infsup(-1,1);
>> Y = infsup(1/2,5/2);
>> qdist(X,Y)
ans = 1.5000
```

Similarly, the width of an interval `X` is implemented with the function `diam(X)`. (For a list of these and other INTLAB functions, see Appendix E. Help for a specific function listed there can be obtained in the command window with the "help" command. For example, issuing `help diam` gives a summary of how to use the INTLAB `diam` function.[18]) Illustrate Exercise 6.4 by using the INTLAB functions `qdist` and `diam`. □

6.2 Refinement

According to the fundamental theorem of interval analysis, we have

$$f(X_1, \ldots, X_n) \subseteq F(X_1, \ldots, X_n) \tag{6.4}$$

for any inclusion isotonic interval extension F of a real function f. This means that an interval value of F contains the range of values of the corresponding real function f when the real arguments of f lie in the intervals $X_1, \ldots, X_n$. In Chapter 5, we saw an example in which equality held in (6.4), but we also saw examples in which the width of the right-hand side substantially exceeded that of the left-hand side.

Lipschitz Interval Extensions

We now examine a process by which $f(X_1, \ldots, X_n)$ may be approximated as closely as desired by a finite union of intervals. We begin with a concept closely related to continuity.

Definition 6.1. An interval extension F is said to be *Lipschitz in* X_0 if there is a constant L such that $w(F(X)) \le Lw(X)$ for every $X \subseteq X_0$.

Hence, the width of $F(X)$ approaches zero at least linearly with the width of X. Here X may be an interval or an interval vector $X = (X_1, \ldots, X_n)$.

Lemma 6.1. If F is a natural interval extension of a real rational function with $F(X)$ defined for $X \subseteq X_0$, where X and X_0 are intervals or n-dimensional interval vectors, then F is Lipschitz in X_0.

[18]Issuing the `help` command for a particular function will also list, in blue, different versions of the function, for different data types. Clicking on one of these links gives information about the function when used with that data type.

Proof. For any real numbers a and b and any intervals X_i and Y_i, we have the following relations (which are not hard to prove):

$$\begin{aligned} w(aX_i + bY_i) &= |a|w(X_i) + |b|w(Y_i), \\ w(X_iY_i) &\le |X_i|w(Y_i) + |Y_i|w(X_i), \\ w(1/Y_i) &\le |1/Y_i|^2 w(Y_i) \text{ if } 0 \notin Y_i. \end{aligned} \tag{6.5}$$

Since the natural interval extension has interval values $F(X)$ obtained by a fixed finite sequence of interval arithmetic operations on real constants (from a given finite set of coefficients) and on the components of X (if X is an interval vector) and since $X \subseteq X_0$ implies that $|X_i| \le \|X_0\|$ for every component of X, it follows that a finite number of applications of (6.5) will produce a constant L such that $w(F(X)) \le Lw(X)$ for all $X \subseteq X_0$. □

By the following result, certain interval extensions of irrational functions are also Lipschitz.

Lemma 6.2. If a real-valued function $f(x)$ satisfies an ordinary Lipschitz condition in X_0,

$$|f(x) - f(y)| \le L|x - y| \quad \text{for } x, y \in X_0,$$

then the united extension of f is a Lipschitz interval extension in X_0.

Proof. The function f is necessarily continuous. The interval (or interval vector) X_0 is compact. Thus, $w(f(X)) = |f(x_1) - f(x_2)|$ for some $x_1, x_2 \in X \subseteq X_0$. But $|x_1 - x_2| \le w(X)$; therefore, $w(f(X)) \le Lw(X)$ for $X \subseteq X_0$. □

It follows that interval extensions which are united extensions, such as the following, are also Lipschitz in X_0:

1. X^n given by (5.29);
2. $e^X = [e^{\underline{X}}, e^{\overline{X}}]$;
3. $X^{1/2} = [\underline{X}^{1/2}, \overline{X}^{1/2}]$ for $0 < \underline{X}_0$;
4. $\ln X = [\ln \underline{X}, \ln \overline{X}]$ for $0 < \underline{X}_0$;
5. $\sin X = [\sin \underline{X}, \sin \overline{X}]$ for $X_0 \subseteq [-\frac{\pi}{2}, \frac{\pi}{2}]$;
6. $\sin X = [\min_{x \in X}(\sin x), \max_{x \in X}(\sin x)]$ for arbitrary X_0.

Lemma 6.3. Let F and G be inclusion isotonic interval extensions with F Lipschitz in Y_0, G Lipschitz in X_0, and $G(X_0) \subseteq Y_0$. Then the composition $H(X) = F(G(X))$ is Lipschitz in X_0 and is inclusion isotonic.

Proof. The inequality

$$w(H(X)) = w(F(G(X)) \le L_2 w(G(X)) \le L_2 L_1 w(X)$$

shows that H is Lipschitz in X_0. □

Example 6.2. If $G(X)$ is rational in X and $G(X_0)$ is defined, then $\sin G(X)$ is Lipschitz and inclusion isotonic for $X \subseteq X_0$. Ordinary monotonicity is *not* required for inclusion isotonicity. □

Thus, an inclusion isotonic interval extension whose values $F(X)$ are defined by a fixed finite sequence of rational (interval arithmetic) operations or compositions of Lipschitz extensions is a Lipschitz extension in some suitably chosen region.

Subdivisions and Refinements

We are now in a position to accomplish the goal of this section.

Definition 6.2. By a *uniform subdivision* of an interval vector $X = (X_1, \ldots, X_n)$, we mean the following. Let N be a positive integer and define

$$X_{i,j} = \left[\underline{X}_i + (j-1)\, w(X_i)/N\,,\ \underline{X}_i + j\, w(X_i)/N\right], \qquad j = 1, \ldots, N.$$

We have $X_i = \cup_{j=1}^{N} X_{i,j}$ and $w(X_{i,j}) = w(X_i)/N$. Furthermore,

$$X = \cup_{j_i=1}^{N} (X_{1,j_1}, \ldots, X_{n,j_n}), \tag{6.6}$$

with $w(X_{1,j_1}, \ldots, X_{n,j_n}) = w(X)/N$.

Definition 6.3. Let $F(X)$ be an inclusion isotonic, Lipschitz, interval extension for $X \subseteq X_0$. The interval quantity

$$F_{(N)}(X) = \bigcup_{j_i=1}^{N} F(X_{1,j_1}, \ldots, X_{n,j_n}) \tag{6.7}$$

is called a *refinement* of F over X.

A refinement $F_{(N)}(X)$ of $F(X)$ is the union of the interval values of F over the elements of a uniform subdivision of X.

If X and Y are intervals such that $X \subseteq Y$, then there is an interval E with $\underline{E} \leq 0 \leq \overline{E}$ such that $Y = X + E$ and $w(Y) = w(X) + w(E)$. If F is an inclusion isotonic interval extension of f with $F(X)$ defined for $X \subseteq X_0$, then $f(X) \subseteq F(X)$ for $X \subseteq X_0$. We have $F(X) = f(X) + E(X)$ for some interval-valued function $E(X)$ with $w(F(X)) = w(f(X)) + w(E(X))$.

Definition 6.4. We call

$$w(E(X)) = w(F(X)) - w(f(X)) \tag{6.8}$$

the *excess width* of $F(X)$.

The united extension itself has zero excess width; $f(X)$ gives the exact range of values of a continuous function $f(x)$ for $x \in X$. We can compute $f(X)$ as closely as desired by computing refinements of an extension $F(X)$.

Theorem 6.1. If $F(X)$ is an inclusion isotonic, Lipschitz, interval extension for $X \subseteq X_0$, then the excess width of a refinement $F_{(N)}(X)$ is of order $1/N$. We have

$$F_{(N)}(X) = f(X_1, \ldots, X_n) + E_N, \tag{6.9}$$

where $w(E_N) \leq K w(X)/N$ for some constant K.

Proof. Relation (6.9) follows from the fact that $f(X) = \cup_s f(X_s)$, where the X_s are the elements of the uniform subdivision. We have $F(X_s) = f(X_s) + E_s$ for some E_s and

$$w(E_s) = w(F(X_s)) - w(f(X_s)) \leq w(F(X_s)) \leq Lw(X_s) \leq Lw(X)/N$$

for each X_s. The inequality for $w(E_N)$ holds with $K = 2L$ since, in the worst case, the maximum excess width may have to be added to both upper and lower bounds in the union. □

We can compute arbitrarily sharp lower and upper bounds on the exact range of values of a wide variety of real-valued functions of n variables by subdividing the domain of arguments and taking the union of interval evaluations over the elements of the subdivision. This technique, also known as *splitting*, converges at least linearly in the width of the pieces in the subdivision.

Example 6.3. Consider the function $f(x) = x - x^2$ for $x \in [0, 1]$. It is easily verified that $f([0, 1]) = \left[0, \frac{1}{4}\right]$. The interval extension

$$F(X) = X - X \cdot X$$

has $F([0, 1]) = [-1, 1]$. Hence, its excess width is $2 - \frac{1}{4} = \frac{7}{4}$. Let us split $[0, 1]$ into n subintervals $X_i = [(i-1)/n, i/n]$, $1 \leq i \leq n$. We now compute the range of values of $F(X)$ over each of these subintervals and then take the union of the resulting intervals to find a narrower interval still containing the range of values of $F(X)$. The table below gives numerical results for a few values of n:

n	$F_{(n)}(X)$
1	$[-1, 1]$
2	$[-0.5, 0.75]$
10	$[-0.1, 0.35]$
100	$[-0.01, 0.26]$
1000	$[-0.001, 0.251]$
10000	$[-0.0001, 0.2501]$

These strongly suggest convergence, and every $F_{(n)}(X)$ contains the exact range of values $\left[0, \frac{1}{4}\right]$. □

Exercise 6.6. The computations in Example 6.3 can be done neatly in MATLAB using the INTLAB toolbox. In particular, $F(X)$ can be programmed:

```
function Y = example6p3(X)
Y = X - X*X;
```

(The above MATLAB code should be put into a file named `example6p3.m`. This happens automatically if MATLAB's editor is used and the default "save file" is done the first time the file is saved.) Computation of a row of the table in Example 6.3 can be done with the following function:

```
function Y = refinement(X,f,N)
% Y = refinement(X,f,N)
```

```
% computes a uniform refinement of the interval X with N
% subintervals, to compute an interval enclosure for
% the range of the interval function f (passed as a
% character string).  See Section 6.2 of "Introduction
% to Interval Analysis" by Moore, Kearfott, and Cloud.

% First form the N subintervals --
h = (sup(X)-inf(X))/N;
xi=inf(X);
x1=xi;
for i=1:N
   % This is more accurate for large N than
   % xip1 = xi + h
    xip1 = x1 + i*h;
    Xs(i) =infsup(xi,xip1);
   % Do it this way so there are no "cracks" due
   % to roundoff error --
    xi=xip1;
end
% Redefine the upper bound to eliminate roundoff
% error problems --
Xs(N) = infsup(inf(Xs(N)),sup(X));

% Now compute the extension by computing the natural
% extensions over the subintervals and taking the union --
Y = feval(f,Xs(1));
if N>1
    for i=2:N
        Y = hull(Y,feval(f,Xs(i)));
    end
end
```

For example, we might create the following dialogue in the MATLAB command window:

```
>> X = infsup(0,1);
>> Y = refinement(X,'example6p3',10)
intval Y = [ -0.1000 , 0.3501 ]
```

Use this to check the table in Example 6.3. □

6.3 Finite Convergence and Stopping Criteria

Nested Interval Sequences

Definition 6.5. An interval sequence $\{X_k\}$ is *nested* if $X_{k+1} \subseteq X_k$ for all k.

Lemma 6.4. Every nested sequence $\{X_k\}$ converges and has the limit $\cap_{k=1}^{\infty} X_k$.

Proof. $\{\underline{X}_k\}$ is a nondecreasing sequence of real numbers, bounded above by $\overline{X}_1$, and so has a limit $\underline{X}$. Similarly, $\{\overline{X}_k\}$ is nonincreasing and bounded below by $\underline{X}_1$ and so has a limit $\overline{X}$. Furthermore, since $\underline{X}_k \leq \overline{X}_k$ for all k, we have $\underline{X} \leq \overline{X}$. Thus $\{X_k\}$ converges to $X = [\underline{X}, \overline{X}] = \cap_{k=1}^{\infty} X_k$. □

Lemma 6.5. Suppose $\{X_k\}$ is such that there is a real number $x \in X_k$ for all k. Define $\{Y_k\}$ by $Y_1 = X_1$ and $Y_{k+1} = X_{k+1} \cap Y_k$ for $k = 1, 2, \ldots$. Then Y_k is nested with limit Y, and

$$x \in Y \subseteq Y_k \text{ for all } k. \tag{6.10}$$

Proof. By induction, the intersection defining Y_{k+1} is nonempty so $\{Y_k\}$ is well defined. It is nested by construction. Relation (6.10) follows from Lemma 6.4. □

Finite Convergence

Definition 6.6. By the *finite convergence* of a sequence $\{X_k\}$, we mean there is a positive integer K such that $X_k = X_K$ for $k \geq K$. Such a sequence *converges in K steps*.

Example 6.4. It is not hard to see that

$$X_0 = [1, 2], \qquad X_{k+1} = 1 + X_k/3 \quad (k = 0, 1, 2, \ldots),$$

generates a nested sequence $\{X_k\}$. The rational interval function $F(X) = 1 + X/3$ is inclusion isotonic. Therefore, $X_1 = F(X_0) = 1 + [1, 2]/3 = \left[\frac{4}{3}, \frac{5}{3}\right] \subseteq X_0 = [1, 2]$. It follows that $X_{k+1} = F(X_k) \subseteq X_k$ for all k by finite induction. By Lemma 6.4, the sequence has a limit X. If we compute $\{X_k\}$ using IA, we will obtain a sequence $\{X_k^*\}$ with $X_k \subseteq X_k^*$ for all k. More precisely, let X_k^* be defined by $X_0^* = X_0 = [1, 2]$, and

$$X_{k+1}^* = \{1 + X_k^*/3 \text{: computed in IA}\} \cap X_k^*$$

for $k = 0, 1, 2, \ldots$ (which we can apply, since $X \subseteq X_k$ for all k). It follows from Lemma 6.5 that X_k^* is nested and that the limit of $\{X_k\}$ is contained in the limit X^* of $\{X_k^*\}$. The sequence $\{X_k^*\}$ will converge in a finite number of steps. For instance, with three-digit IA we find

$$\begin{aligned}
X_0^* &= [1, 2],\\
X_1^* &= [1.33, 1.67],\\
X_2^* &= [1.44, 1.56],\\
X_3^* &= [1.48, 1.52],\\
X_4^* &= [1.49, 1.51],\\
X_5^* &= [1.49, 1.51],
\end{aligned}$$

and $X_k^* = X^*$ for all $k \geq 4$. We have finite convergence in four steps. The real sequence $x_{k+1} = 1 + x_k/3$ converges to 1.50 (after infinitely many steps) from any x_0. □

We can also illustrate the finite convergence of the sequence in Example 6.4, as well as illustrate use of the uncertainty representation within INTLAB, with the following INTLAB function:

```
function [X] = finite_convergence_example()
% Implements the computations for the finite convergence
% example in Chapter 6 of Moore / Kearfott / Cloud
format long;
intvalinit('Display_');
```

```
X = infsup(1,2);
X_new = infsup(-Inf,Inf);
i=0;
while (X_new ~= X)
    i=i+1;
    X = intersect(X,X_new);
    i,X
    X_new = 1+ X/3;
end
```

When finite_convergence_example is issued from the MATLAB command prompt, the following output (edited for brevity) ensues:

```
finite_convergence_example
===> Default display of intervals with uncertainty (e.g. 3.14_),
         inf/sup or mid/rad if input too wide
i =      1,  intval X = [ 1.00000000000000 , 2.00000000000000 ]
i =      2,  intval X = [ 1.33333333333333 , 1.66666666666667 ]
i =      3,  intval X = [ 1.44444444444444 , 1.55555555555556 ]
i =      4,  intval X =   1.5_____________
i =      5,  intval X =   1.50____________
i =      6,  intval X =   1.50____________
i =      7,  intval X =   1.500___________
i =      8,  intval X =   1.500___________
                      .
                      .
                      .
i =     28,  intval X =   1.5000000000000_
i =     29,  intval X =   1.5000000000000_
i =     30,  intval X =   1.50000000000000
i =     31,  intval X =   1.50000000000000
i =     32,  intval X =   1.50000000000000
i =     33,  intval X =   1.50000000000000
i =     34,  intval X =   1.50000000000000
diary off
```

This illustrates that approximately half a decimal point of accuracy is gained on each iteration and that, indeed, the convergence ends after a finite number of steps (34 steps) when using standard binary arithmetic.

Next we examine a sequence of refinements computed in IA and intersected to yield a nested sequence that converges in a finite number of steps.

Example 6.5. Consider the interval polynomial $F(X) = X(1 - X)$ and its refinements $F_{(N)}(X)$. Let $X = [0, 1]$ and take the sequence

$$Y_1 = F_{(1)}([0,1]) = F([0,1]), \qquad Y_{k+1} = F_{(k+1)}([0,1]) \cap Y_k \quad (k = 1, 2, \ldots).$$

The intersections are nonempty because each refinement contains the range of values of $f(x) = x(1 - x)$ for $x \in [0, 1]$, namely $[0, \frac{1}{4}]$. By construction, $\{Y_k\}$ is nested. By Theorem 6.1 and Lemma 6.5,

$$f(X) = \bigcap_{k=1}^{\infty} F_{(k)}(X) = \lim_{k\to\infty} Y_k. \tag{6.11}$$

We have

$$F_{(k)}([0,1]) = \begin{cases} \left[0, \frac{1}{4} + 1/(2k) + 1/(4k^2)\right], & k \text{ odd}, \\ \left[0, \frac{1}{4} + 1/(2k)\right], & k \text{ even}. \end{cases}$$

We can compute a subsequence of $\{Y_k\}$, for example, for k of the form $k = 2^t$, $t = 1, 2, \ldots$. If we compute the nested sequence of intervals $Y_1^* = [0, 1]$, and for $t = 1, 2, \ldots$,

$$Y_{t+1}^* = F_{(2^t)}^*([0,1]) \cap Y_i^*,$$

where $F_{(k)}^*([0,1])$ is $F_{(k)}([0,1])$ computed in three-place IA, we obtain again a nested sequence $\{Y_t^*\}$. This converges in nine steps to the interval $[0, 0.251]$. We have $Y_t^* = [0, 0.251]$ for all $t \geq 9$. □

Natural Stopping Criterion

For any fixed (as opposed to "variable") precision representation of machine numbers, there is a finite set of machine numbers represented by strings of bits $b_0 b_1 \ldots b_s$ with s fixed. Hence there is only a finite set of intervals with machine number endpoints. *Any nested sequence of such intervals is necessarily finitely convergent.*

For any iterative interval method that produces a nested sequence with endpoints represented by fixed precision machine numbers, we have a natural stopping criterion. Since the sequence $\{X_k\}$ converges in a finite number of steps, we can compute the X_k until

$$X_{k+1} = X_k. \tag{6.12}$$

If the X_k are generated by a procedure of the form

$$X_{k+1} = F(X_k) \tag{6.13}$$

such that each X_{k+1} depends only on the previous X_k, then it is clear that (6.12) is sufficient to guarantee convergence.

In particular, if $F(X)$ is a rational expression in X and if X_0 is an interval such that $F(X_0) \subseteq X_0$, which can be tested on the computer, it follows that $\{X_k\}$ defined by

$$X_{k+1} = F(X_k) \qquad (k = 0, 1, 2, \ldots) \tag{6.14}$$

is nested with

$$X_0 \supseteq X_1 \supseteq X_2 \supseteq \cdots$$

and hence converges to some X^* with $X^* = F(X^*)$ and $X^* \subseteq X_k$ for all $k = 0, 1, 2, \ldots$.

With IA, it may happen that $X_1 = F(X_0) \subseteq X_0$ but that $X_{k+1} \not\subseteq X_k$ for some k. If instead of (6.14), we compute

$$X_{k+1} = F(X_k) \cap X_k \tag{6.15}$$

and stop when (6.12) is satisfied (which will happen always for some k using fixed finite precision IA), we have the narrowest possible interval containing $X^* = F(X^*)$. A narrower interval would require using higher-precision arithmetic.

Furthermore, if for a chosen X_0, the interval $F(X_0) \cap X_0$ is empty, then X_0 contains no fixed points of F (i.e., there is no $X \in X_0$ such that $F(X) = X$). This follows from the inclusion isotonicity of F, for if $F(X) = X$ and $X \subseteq X_0$, then $X = F(X) \subseteq F(X_0)$ and so $X \subseteq F(X_0) \cap X_0$; therefore, if $F(X_0) \cap X_0$ is empty, there is no such X.

Example 6.6. Let

$$F(X) = \tfrac{1}{2}X + 2.$$

If we take $X_0 = [1, 2]$, then

$$F(X_0) = \tfrac{1}{2}[1, 2] + 2 = \left[\tfrac{5}{2}, 3\right].$$

Since $F(X_0) \cap X_0 = \emptyset$, there is no fixed point of F in $[1, 2]$. If we take $X_0 = \left[2, \tfrac{7}{2}\right]$ instead, then

$$F(X_0) = \tfrac{1}{2}\left[2, \tfrac{7}{2}\right] + 2 = \left[3, \tfrac{15}{4}\right].$$

Here $F(X_0) \cap X_0 = \left[3, \tfrac{7}{2}\right]$; we cannot conclude anything since $F(X_0) \not\subseteq X_0$. Finally, with $X_0 = [2, 5]$, we have $F(X_0) = F([2, 5]) = \left[3, \tfrac{9}{2}\right]$. This time, $F(X_0) \subseteq X_0$, so F has a fixed point X in X_0. The iterations

$$X_{k+1} = F(X_k) \cap X_k, \qquad k = 0, 1, 2, \dots, \tag{6.16}$$

produce, in three-digit IA,

$$\begin{aligned}
X_1 &= [3, 4.5] \cap [2, 5] = [3, 4.5],\\
X_2 &= [3.5, 4.25] \cap [3, 4.5] = [3.5, 4.25],\\
X_3 &= [3.75, 4.13] \cap [3.5, 4.25] = [3.75, 4.13],\\
X_4 &= [3.87, 4.07],\\
X_5 &= [3.93, 4.04],\\
X_6 &= [3.96, 4.02],\\
X_7 &= [3.98, 4.01],\\
X_8 &= [3.99, 4.01],\\
X_9 &= [3.99, 4.01],\\
X_{k+1} &= X_k, \quad k = 8, 9, 10, \dots.
\end{aligned}$$

F has a fixed point in $[3.99, 4.01]$. □

If the process generating the sequence depends explicitly on k as well as on X_k, say,

$$X_{k+1} = F(k, X_k),$$

then we might have $X_{k+1} = X_k$ for some k and yet $X_{k+2} \neq X_k$ even though $\{X_k\}$ is nested. An example is

$$X_{k+1} = ([0, 2]/k) \cap X_k, \qquad X_1 = [0, 1]. \tag{6.17}$$

Here

$$\begin{aligned} X_1 &= [0, 1], \\ X_2 &= [0, 1], \\ X_3 &= [0, 1], \\ X_4 &= [0, \tfrac{2}{3}], \\ X_5 &= [0, \tfrac{1}{2}], \\ &\vdots \\ X_{k+1} &= [0, 2/k], \quad k > 2. \end{aligned}$$

Hence, (6.12) is a valid stopping criterion if and only if $\{X_k\}$ is nested and generated by (6.16) with $F(X_k)$ depending only on X_k.

Remarks on Branching

Let us close this section with some comments about branching in computer programs. For real numbers x and y, the relation $x \le y$ is either true or false. If it is false, then we have $x > y$. If all we know about x and y is that they lie in intervals $x \in X$ and $y \in Y$, then we can only deduce one of three possibilities:

(1) If $\overline{X} \le \underline{Y}$, then $x \le y$.

(2) If $\underline{X} > \overline{Y}$, then $x > y$.

(3) Otherwise, we *don't know* whether $x \le y$ or $x > y$.

We can modify any computer program by using IA instead of ordinary machine arithmetic whenever rounding error is possible in an arithmetic operation and by using the above three-valued logic for all branch tests involving computed quantities which are only known to lie in certain computed intervals. If we stop the computation whenever a "don't know" logical value is obtained, then all intervals computed up to that point by the modified problem contain the exact result of the corresponding finite sequence of operations using exact (infinite precision) real arithmetic for all arithmetic operations.

Note Although we may, in principle, apply this to all computer programs originally written with floating point arithmetic, it is unlikely to lead to good results without additional careful analysis and redesign. For example, if we "don't know" whether $x \le y$ or $x > y$, we must take the union of the values of the separate branches, resulting in more computation and wider intervals. This is often impractical, especially if there are many such branches, and can lead to unusefully wide enclosures.

Exercise 6.7. The following MATLAB function (which uses INTLAB) performs a single step of the iteration (6.15):

```
function new_X = interval_fixed_point_step(f,X)
% X_star = interval_fixed_point_step(f,X)
```

```
% returns the result of a step of interval
% fixed point iteration
% X_{k+1} = intersect(X_k, f(X_k))  as explained
% in Section 6.3.

new_X = intersect(X,feval(f,X));
```

For example, with the function

```
function Y = example6p4(X)
Y = 1 + X/3;
```

we can reproduce an analogue of the sequence in Example 6.4, but with correctly outwardly rounded intervals displayed roughly to machine precision, with the following MATLAB dialogue:

```
>> format long
>> intvalinit('DisplayInfsup')
>> X = infsup(1,2)
intval X = [ 1.00000000000000 , 2.00000000000000 ]
>> X = interval_fixed_point_step('example6p4',X)
intval X = [ 1.33333333333333 , 1.66666666666667 ]
>> X = interval_fixed_point_step('example6p4',X)
intval X = [ 1.44444444444444 , 1.55555555555556 ]
>> X = interval_fixed_point_step('example6p4',X)
intval X = [ 1.48148148148148 , 1.51851851851852 ]
>> X = interval_fixed_point_step('example6p4',X)
intval X = [ 1.49382716049382 , 1.50617283950618 ]
>> X = interval_fixed_point_step('example6p4',X)
intval X = [ 1.49794238683127 , 1.50205761316873 ]
```

Do the same iteration for Example 6.6. □

Exercise 6.8. The MATLAB function from Exercise 6.7 implementing a step of interval fixed point iteration can be called iteratively to convergence with the following MATLAB function:

```
function [X_limit,N] = interval_fixed_point_limit(f,X)
% [X_limit,N] = interval_fixed_point_convergence(f,X) returns
% the limit of fixed point iteration X_{k+1} = f(X k) as
% explained in Section 6.3, and returns the number of iterations
% required in N.  If one or both of the bounds of X_limit is NaN,
% then the limit is the empty set.
old_X = infsup(-Inf,Inf);
N = 0;
while (old_X ~= X) & ~( isnan(inf(X)) | isnan(sup(X)) )
    N = N+1;
    old_X = X;
    X = interval_fixed_point_step(f,X);
end
X_limit = X;
```

For instance, we might have the following MATLAB dialogue for Example 6.4:

```
>> X = infsup(1,2)
intval X = [ 1.00000000000000 , 2.00000000000000 ]
>> [Y,N] = interval_fixed_point_limit('example6p4',X)
intval Y = [ 1.49999999999999 , 1.50000000000001 ]
N = 34
```

(The displayed interval `[1.49999999999999, 1.50000000000001]` represents an outwardly rounded binary-to-decimal conversion of the actual binary representation of the interval.)

1. Does `interval_fixed_point_limit` work for Examples 6.5 and 6.6?
2. Can `interval_fixed_point_limit` fail? If so, how?
3. Do you think an upper bound on the number of iterations allowed in the loop in `interval_fixed_point_limit` should be given? If not, then, assuming IEEE double precision arithmetic, can you compute an upper bound on the number of iterations actually needed? □

6.4 More Efficient Refinements

We have seen that refinement of an interval extension is a method for computing arbitrarily sharp upper and lower bounds on the range of values of a real function, but it is of practical importance to be able to obtain these bounds as efficiently as possible. We begin by characterizing the class of functions most commonly used in computing.

Code Lists and the Class FC

Let $\mathrm{FC}_n(X_0)$ be the class of real-valued functions f of n real variables whose values $f(x_1, \dots, x_n)$ for each f are defined for all $x = (x_1, \dots, x_n)$ in the interval vector $X_0 = (X_{10}, \dots, X_{n0})$ by some fixed, finite sequence of operations

$$
\begin{aligned}
T_1 &= h_1(y_1, z_1) \\
T_2 &= h_2(y_2, z_2) \\
&\vdots \\
f(x) = T_M &= h_M(y_m, z_m)
\end{aligned}
\tag{6.18}
$$

for $y_1, z_1 \in S_1 = \{x_1, \dots, x_n, c_1, \dots, c_p\}$, where $c_1, \dots, c_p$ are given real numbers (the *coefficients* of f), and where $h_1, \dots, h_M$ are the arithmetic functions $+, -, \cdot, /$, or *unary* functions (of only one of the variables y_i, z_i) of elementary type: $\exp(\cdot)$, $\ln(\cdot)$, $\sqrt{\cdot}$, etc. Furthermore, $y_{i+1}, z_{i+1} \in S_{i+1} = S_i \cup \{T_i\}$, $i = 1, \dots, M-1$. The integer M, the coefficients $c_1, \dots, c_p$ (if any), and the sequence of operations $h_1, \dots, h_M$ define a particular function $f \in \mathrm{FC}_n(X_0)$. Each y_i, z_i may be an x; or a constant c_i or one of the previous T_j $(j < i)$.

Example 6.7. The function f defined by

$$f(x_1, x_2) = x_1 e^{x_1 + x_2^2} - x_2^2$$

can be expressed as the sequence (commonly called a *code list* or *computational graph*)

$$\begin{aligned} T_1 &= x_2^2, \\ T_2 &= x_1 + T_1, \\ T_3 &= e^{T_2}, \\ T_4 &= T_3 x_1, \\ f(x_1, x_2) &= T_4 - T_1. \end{aligned}$$

Therefore, it belongs to $\mathrm{FC}_2(X_0)$ for any X_0. □

We assume that we can compute arbitrarily sharp bounds on the exact range of values of each of the unary functions occurring in the sequence of operations for $f(x)$. The class FC_n is further assumed to include only functions f that are defined for all $x \in X_0$ and that have inclusion isotonic, Lipschitz, interval extensions $F(X)$ for $X \subseteq X_0$. For some of the discussion to follow, we also assume that the first and second partial derivatives of f satisfy the same conditions as f. This final assumption, concerning the existence of interval extensions with desirable properties (inclusion isotonicity, etc.) is satisfied automatically for all $f \in \mathrm{FC}_n(X_0)$, provided only that $F(X_0)$ is defined, where $F(X)$ is the interval extension

$$\begin{aligned} T_1 &= H_1(Y_1, Z_1), \\ T_2 &= H_2(Y_2, Z_2), \\ &\vdots \\ F(x) = T_M &= H_M(Y_m, Z_m), \end{aligned} \tag{6.19}$$

where $H_1, \ldots, H_M$ are the united extensions of the functions $h_1, \ldots, h_M$, respectively. This is the case, replacing real arithmetic operations by interval arithmetic operations and using the united extensions of the unary functions, as long as for the interval vector X_0, no division by an interval containing 0 occurs during the computation of $T_1, \ldots, T_M$ and provided that no arguments containing zero occur for such unary operations as $\ln(\cdot)$ or $\sqrt{\cdot}$. (The latter would spoil the Lipschitz property for the interval extension.) Since such an F is inclusion isotonic, $F(X)$ is defined for every $X \subseteq X_0$.

Under these conditions, Theorem 6.1 states that the excess width of the refinement $F_{(N)}(X)$ of the interval extension $F(X)$ given by (6.19) of a function $f \in \mathrm{FC}_n(X_0)$ satisfies

$$w(E_n) \le K_f w(X)/N \text{ for some } K_f \text{ depending only on } f. \tag{6.20}$$

An evaluation of F requires one pass through the finite sequence of operations (6.19), just as an evaluation of f requires one pass through (6.18). An evaluation of $F_{(N)}(X)$, given by (6.9), requires N^n evaluations of $F(X)$. If n is at all large, this involves a prohibitive amount of computation to achieve the result (6.20) for large N. Even for $n = 2$ we have, for $N = 1000$ (perhaps to reduce the excess width to 0.001), $1000^2 = 10^6$ evaluations to carry out.

Fortunately, by using more information about a particular function f, we can compute an interval containing the exact range of values and having arbitrarily small excess width with far less work. In some cases, we can even compute an interval with zero excess width in one evaluation of $F(X)$. In the remainder of this section, we assume that $f \in \mathrm{FC}_n(X_0)$ with $F(X)$ defined for $X \subseteq X_0$ by (6.19).

Theorem 6.2. If each of the variables $x_1, \ldots, x_n$ occurs at most once in the list $y_1, z_1, \ldots, y_n, z_n$ of arguments in the sequence (6.18) defining $f(x)$, then the interval extension $F(X)$ defined by (6.19) is the united extension of f for all $X \subseteq X_0$.

Proof. We have $F(X) = f(X)$ if and only if $f(X) \subseteq F(X)$ and $F(X) \subseteq f(X)$. The first inclusion holds under our assumptions on $F(X)$. Under the further hypothesis of the theorem, every real number $r \in F(X)$ is expressible as $r = f(x)$ for some $x = (x_1, \ldots, x_n)$ with $x_i \in X_i$ for $i = 1, \ldots, n$. In fact, each H_i is the united extension of h_i. Therefore $F(X) \subseteq f(X)$. ☐

Hence, for a function f satisfying the hypotheses of Theorem 6.2, we can compute the exact range of values of $f(x)$ for $x \in X \subseteq X_0$ with one evaluation of $F(X)$ using (6.19).

If some x_i occurs more than once in the list of arguments in (6.18), then the corresponding X_i occurs more than once in the list of arguments in (6.19). In this case, there may be real numbers in $F(X)$ which are not expressible as $f(x)$ for any $x \in X$.

Example 6.8. Consider $f \in \mathrm{FC}_1([0, 1])$ defined by

$$T_1 = 1 - x_1, \qquad f(x_1) = T_2 = T_1 x_1.$$

The variable x_1 occurs twice as an argument. The corresponding list for $F(X)$ is

$$T_1 = 1 - X_1, \qquad F(X_1) = T_2 = T_1 X_1.$$

Now $F([0, 1]) = [0, 1]$, but there is no $x_1 \in [0, 1]$ such that $f(x_1) = (1 - x_1)x_1 = 1 \in [0, 1]$. On the other hand, the interval function

$$T_1 = 1 - X_1, \qquad F(X_1, X_2) = T_2 = T_1 X_2,$$

is the united extension of the real function

$$T_1 = 1 - x_1, \qquad f(x_1, x_2) = T_2 = T_1 x_2.$$

In particular, $F([0, 1], [0, 1]) = [0, 1]$ is the exact range of values of $f(x_1, x_2) = (1 - x_1)x_2$ for $x_1 \in [0, 1]$ and $x_2 \in [0, 1]$.

Another explanation for the difference between these two results is the following. In both cases, we first compute $1 - [0, 1] = [0, 1]$. This is the exact range of values of $1 - x_1$ for $x_1 \in [0, 1]$. The discrepancy comes at the next step. When we compute $T_1 X_1$ with $T_1 = [0, 1]$ and $X_1 = [0, 1]$, we obtain $[0, 1]$. We obtain the same result for $T_1 X_2$ with $T_1 = [0, 1]$ and $X_2 = [0, 1]$.

In the second case, this is the desired exact range of values of $f(x_1, x_2)$. In the first case, we get an overestimation of the range of values of $f(x_1) = (1 - x_1)x_1$ because, during the numerical evaluation of $T_1 X_1$, we have not made use of the information that $T_1 = [0, 1]$ represents the range of values of $1 - x_1$ for $x_1 \in X_1$. We have not distinguished between this and $T_1 = [0, 1]$ as the range of values of $1 - x_2$ for a variable x_2 independent of x_1. ☐

Example 6.8 shows that it is better to use the interval extension X^n defined by (5.29) for integer powers x^n than to use $X \cdot X \cdot \cdots \cdot X$. For $X = [-1, 2]$ we obtain $X \cdot X = [-2, 4]$, whereas X^2 using (5.29) yields the correct result, $[-1, 2]^2 = [0, 4] = \{x^2 \colon x \in [-1, 2]\}$.

Exercise 6.9. Find the range of values of

$$g(x, y, z, u, v, w) = (33 - 2x + yz)/(10 - uvw)$$

over the unit 6-cube $x, y, z, u, v, w \in [0, 1]$. □

Sometimes we can reduce the number of occurrences of a variable in a real rational expression by algebraic manipulation.

Example 6.9. Let $f(x_1, x_2, x_3)$ be defined by

$$f(x_1, x_2, x_3) = \frac{x_1 + x_2}{x_1 - x_2} x_3, \tag{6.21}$$

which we can rewrite as

$$f(x_1, x_2, x_3) = x_3 \left(1 + \frac{2}{(x_1/x_2) - 1} \right). \tag{6.22}$$

For $x_1 \in [1, 2]$, $x_2 \in [5, 10]$, and $x_3 \in [2, 3]$, the natural interval extension of (6.22) produces the correct range of values $\left[-7, -\frac{22}{9}\right]$ since each variable occurs only once. However, the natural interval extension of (6.21) produces $\left[-12, -\frac{12}{9}\right]$ with an excess width of $\frac{55}{9}$. □

Given a function $f \in \mathrm{FC}_n(X_0)$ defined by a finite sequence of the form (6.18), we can construct interval extensions of f other than the natural extension given by (6.19). We will discuss four: the *centered form*, the *mean value form*, a variant of the mean value form we call the *slope form*, and the *monotonicity-test form*.

The Centered Form

We now discuss the centered form, a second-order interval extension. To obtain the centered form $F_c(X_1, \ldots, X_n)$, we first rewrite $f(x_1, \ldots, x_n)$ as

$$f(x_1, \ldots, x_n) = f(c_1, \ldots, c_n) + g(y_1, \ldots, y_n) \tag{6.23}$$

with $y_i = x_i - c_i$ and with g defined by (6.23):

$$g(y_1, \ldots, y_n) = f(y_1 + c_1, \ldots, y_n + c_n) - f(c_1, \ldots, c_n).$$

For rational f, we can express g in the form

$$g(y_1, \ldots, y_n) = y_1 h_1(y_1, \ldots, y_n) + \cdots + y_n h_n(y_1, \ldots, y_n),$$

where h_i is rational and $h_i(0, \ldots, 0)$ is defined if $f(c_1, \ldots, c_n)$ is. We define F_c by

$$F_c(X_1, \ldots, X_n) = f(c_1, \ldots, c_n) + \sum_{i=1}^{n} Y_i H_i(Y_1, \ldots, Y_n), \tag{6.24}$$

where $c_i = m(X_i)$, $Y_i = X_i - c_i$, and H_i is the natural interval extension of h_i.

Example 6.10. We reconsider the polynomial $p(x) = 1 - 5x + \frac{1}{3}x^3$ given in (5.25). We rewrite $p(x)$ as

$$p(x) = 1 - 5c + \tfrac{1}{3}c^3 + g(y), \quad \text{where } y = x - c,$$

and

$$g(y) = p(x) - p(c) = y(-5 + \tfrac{1}{3}((y+c)^2 + (y+c)c + c^2)).$$

We define P_c by

$$P_c(X) = p(c) + YH(Y),$$

where $c = m(X)$, $Y = X - m(X)$, and

$$H(Y) = -5 + \tfrac{1}{3}((Y+c)^2 + (Y+c)c + c^2).$$

Now, for $X = [2, 3]$, we obtain

$$c = m(X) = \tfrac{5}{2}, \quad Y = \left[-\tfrac{1}{2}, \tfrac{1}{2}\right], \quad p(c) = p(\tfrac{5}{2}) = -\tfrac{151}{24}.$$

We have $H(Y) = \left[\frac{1}{12}, \frac{31}{12}\right]$, so

$$P_c([2,3]) = \left[-\tfrac{91}{12}, -5\right] = \left[-7.5833333\ldots, -5\right].$$

Compare this with the bounds computed using (5.26) and (5.27). Again, the exact range of values is $[-\frac{10}{3}\sqrt{5} + 1, -5] = [-6.454559\ldots, -5]$. □

From (6.20) it follows (with $N = 1$) that the excess width of the natural interval extension of a rational function is of order $w(X)$. Hansen [63] has shown that the excess width of the centered form is of order $w(X)^2$. Thus, using the centered form in the computation of $F_{(N)}(X)$ defined in (6.7), we have

$$w(E_N) \le K_f w(X)/N^2 \quad \text{for some } K_f$$

instead of (6.20). Using the centered form instead of the natural interval extension for rational functions, we can reduce the excess width of refinements to a prescribed amount with about the square root of the number of function evaluations required for the natural extensions. This often remains prohibitively large, and we will discuss methods for further reduction. Nonetheless, for a single evaluation of an interval extension on an interval vector of small width, the centered form can give very small excess width compared to the natural extension for rational functions. Ratschek [209] has found higher-order centered forms.

Exercise 6.10. Apply the centered form technique to the expression $f(x) = \frac{1}{3}x^3 - 2x$ and the interval $[1.2, 1.6]$. □

Mean Value Form

Another useful interval extension is the mean value form. We can apply this to functions in $FC_n(X_0)$. Let X be an interval vector in X_0 and let $m = m(X)$. Let $D_i F$ be an interval extension of $\partial f/\partial x_i$; we will show how to obtain these presently. By the mean value theorem we have, for all $X \subseteq X_0$,

$$f(X) \subseteq F_{mv}(X) = f(m) + \sum_{i=1}^{n} D_i F(X)(X_i - m_i). \tag{6.25}$$

$F_{mv}(X)$ is the *mean value extension* of f on X.

Example 6.11. Consider

$$f(x_1, x_2) = x_1 e^{x_1 + x_2^2} - x_2^2$$

defined by the sequence

$$\begin{aligned} T_1 &= x_2^2, \\ T_2 &= x_1 + T_1, \\ T_3 &= e^{T_2}, \\ T_4 &= T_3 x_1, \\ f(x) &= T_4 - T_1. \end{aligned} \tag{6.26}$$

Noting that $D_i X_j = 1$ if $j = i$ and 0 if $j \neq i$, we have

$$\begin{aligned} D_1 T_1 &= 0, & D_2 T_1 &= 2X_2, \\ D_1 T_2 &= 1, & D_2 T_2 &= D_2 T_1, \\ D_1 T_3 &= 0, & D_2 T_3 &= T_3(D_2 T_2), \\ D_1 T_4 &= T_3 + X_1(D_1 T_3), & D_2 T_4 &= (D_2 T_3) X_1, \\ D_1 F(X) &= D_1 T_4, & D_2 F(X) &= D_2 T_4 - D_2 T_1. \end{aligned} \tag{6.27}$$

For $X_1 = [1, 2]$ and $X_2 = [0, 1]$, we find $m = (\frac{3}{2}, \frac{1}{2})$, and from (6.26), $f(m) = (3/2)e^{7/4} - \frac{1}{4} = 8.3819040\ldots$. In (6.27), the quantities T_1, T_2, T_3, T_4 must first be computed in interval arithmetic using the natural extensions (or united extensions) of the operations in (6.26) beginning with the given interval values for X_1 and X_2. We find

$$\begin{aligned} T_1 &= [0, 1], \\ T_2 &= [1, 3], \\ T_3 &= [e^1, e^3] = [2.7182818\ldots, 20.0855369\ldots], \\ T_4 &\subseteq [2.7182818, 40.171074]. \end{aligned}$$

We next compute

$$
\begin{aligned}
D_1 T_1 &= 0,\\
D_1 T_2 &= 1,\\
D_1 T_3 &= [2.7182818\ldots, 20.0855369\ldots],\\
D_1 T_4 &\subseteq [5.4365636\ldots, 60.256611],\\
D_1 F(X) &\subseteq [5.4365636, 60.256611],\\
D_2 T_1 &= [0, 2],\\
D_2 T_2 &= [0, 2],\\
D_2 T_3 &\subseteq [0, 40.171074],\\
D_2 T_4 &\subseteq [0, 80.342148],\\
D_2 F(X) &\subseteq [-2, 80.342148].
\end{aligned}
$$

Finally,

$$F_{mv}([1, 2], [0, 1]) \subseteq [-61.917477, 78.681284]$$

by (6.25). □

Exercise 6.11. Repeat Exercise 6.10 using the mean value form. □

The Mean Value Form and INTLAB

The process in Example 6.11 can be fully automated, using a process called *operator overloading*; for example, see [200] for one of the first explanations or [97, section 1.4.4, p. 43ff] for some perspective. Operator overloading consists of extension of the definitions of arithmetic and logical operators and standard functions such as sin, to user-defined data types, such as an interval as an array of two double precision words. Then subroutines are written to implement the arithmetic and logical operations on this data type. In our context, we would have a Taylor data type, and we would compute Taylor coefficients by performing arithmetic operations on this data type. This technique has been implemented in the `gradient` toolbox distributed with INTLAB. While running INTLAB, issuing the command `demo` from within MATLAB's command window opens the demo tab of MATLAB's help screen, from which you can select help on the gradient toolbox and other topics.[19] Computations for Example 6.11 can be done with the following MATLAB command window dialogue:

```
>> format short
>> X = [infsup(1,2), infsup(0,1)]
intval X =
[ 1.0000 , 2.0000 ] [ 0.0000 , 1.0000 ]
>> m = mid(X)
m =
    1.5000    0.5000
>> Xg = gradientinit(X)
intval gradient value Xg.x =
```

[19]Beginning with INTLAB version 5.4, extensive documentation for INTLAB is integrated into MATLAB's help system. In earlier releases of INTLAB, typing `help` in the MATLAB command window gives a list of toolbox help files, including those for INTLAB, upon which one can click.

```
[ 1.0000 , 2.0000 ] [ 0.0000 , 1.0000 ]
intval gradient derivative(s) Xg.dx =
intval Xg.dx(1,1,:) =
[ 1.0000 , 1.0000 ] [ 0.0000 , 0.0000 ]
intval Xg.dx(1,2,:) =
[ 0.0000 , 0.0000 ] [ 1.0000 , 1.0000 ]
>> fm = m(1)*exp(m(1)+m(2)^2) - m(2)^2
fm = 8.3819
>> DFX = Xg(1)*exp(Xg(1)+Xg(2)^2) - Xg(2)^2
intval gradient value DFX.x = [ 1.7182 , 40.1711 ]
intval gradient derivative(s) DFX.dx =
[ 5.4365 , 60.2567 ] [ -2.0000 , 80.3422 ]
>> Xminusm = X-m
intval Xminusm =
[ -0.5000 , 0.5000 ] [ -0.5000 , 0.5000 ]
>> Fmv = fm + DFX.dx * Xminusm'
intval Fmv = [ -61.9175 , 78.6813 ]
```

In fact, using the `gradient` toolbox, we can write the following m-file to compute the mean value form of a general function:

```
function Fmv = mean_value_form(f,X)
% Fmv = mean_value_form(f,X) returns the value for the
% mean value form for f evaluated over the interval X,
% as explained in Section 6.3. If X has more
% than one coordinate, it must be represented as a
% row vector.

m = mid(X);
Xg = gradientinit(X);
fm = feval(f,m);
DFX = feval(f,Xg);
Xminusm = X-m;
Fmv = fm + DFX.dx * Xminusm';
```

If we supply a file `example6p11.m` containing

```
function f = example6p11(x)
f = x(1) * exp(x(1)+x(2)^2) - x(2)^2;
```

then the following MATLAB command window dialogue computes the mean value form corresponding to Example 6.11:

```
>> X = [infsup(1,2), infsup(0,1)];
>> Fmv = mean_value_form('example6p11',X)
intval Fmv = [ -61.9175 , 78.6813 ]
```

Exercise 6.12. Repeat Exercise 6.11 using INTLAB. □

Exercise 6.13. The code list such as (6.27) for Example 6.11 is not unique for a given expression. For example, if $f(x) = x^4 + x^2 + 1$, one sequence might be

$$\begin{aligned} T_1 &= x^2, \\ T_2 &= T_1^2, \\ T_3 &= T_1 + T_2, \\ f(x) &= T_3 + 1, \end{aligned}$$

while another might be

$$\begin{aligned} T_1 &= x^4, \\ T_2 &= x^2, \\ T_3 &= T_2 + 1, \\ f(x) &= T_3 + T_1. \end{aligned}$$

Thus, when operator overloading is used in a compiled language (such as one of the newer Fortran standards or C++) or an interpreted language (such as MATLAB), different code lists might be produced for f with different compilers or different versions of the interpreter. This is usually not of much consequence, unless the compiler also rearranges the expression for "optimization" (typically to reduce the number of operations or to make it possible to evaluate the expression more quickly on particular computer circuitry); as we have seen, rearrangement can result in a different interval enclosure.

(a) Write down two different code lists for

$$f(x) = x_1 e^{x_2} + x_2 e^{2x_2},$$

assuming the system is not rearranged.

(b) Write down a third code list for f, assuming the expression has been rearranged, such as by factoring out an e^{x_2} from both terms.

(c) Evaluate your three code lists step by step using INTLAB. Try different intervals, such as $(X_1, X_2) = ([-1, 1], [-1, 1])$, $(X_1, X_2) = ([-1, 1], [1, 2])$, etc. What do you observe? □

Slope Form

It is possible to replace the interval extensions $D_i F(X)$ in the mean value form (6.25) by intervals $S_i(F, X, m)$ that are often narrower than $D_i F(X)$ but that still result in an enclosure for $f(X)$. The $S_i(F, X, m)$ are interval enclosures for the *slope* of f at m over X.

Definition 6.7. Suppose $f\colon D \subseteq \mathbb{R}^n \to \mathbb{R}$, and suppose $x \in \mathbb{R}^n$, $y \in \mathbb{R}^n$. Then the *slope* $s(f, x, y)$ is defined to be that number such that

$$f(y) - f(x) = s(f, x, y)(y - x).$$

Note that $s(f, x, y)$ exists for particular f, x, and y under various assumptions. For instance, if f has a continuous first derivative, then the mean value theorem states that $s(f, x, y) = f'(\xi)$ for some ξ between x and y.

The power of slopes lies in the fact that interval extensions of them can be computed that are narrower than corresponding interval extensions of derivatives. That is, thinking of x as the variable and y as fixed, we can compute an interval extension $S(f, X, y)$ of $s(f, x, y)$ that is narrower than a corresponding interval extension $F'(X)$.

Example 6.12. If $f(x) = x^2$, then $s(f, x, y) = x + y$. Thus, an interval enclosure $S(f, X, y)$ is $X + y$. For instance, if $X = [1, 3]$ and $y = 2$, then $X + y = [3, 5]$. However, $f'(x) = 2x$, so the range of f' over $X = [1, 3]$ is $2[1, 3] = [2, 6]$, a wider interval than $[3, 5]$. □

For multivariate functions, interval bounds on slopes with respect to particular variables, analogous to interval enclosures for partial derivatives, can be included. We will denote such a bound for the ith variable by $S_i(f, X, y)$. Assuming we have such slope enclosures, we obtain, for $x \in X$ and $Y_i = X_i - y_i$,

$$f(x) \in F_s(X, y) = f(y) + \sum_{i=1}^{n} Y_i S_i(f, X, y). \tag{6.28}$$

We call this the *slope form*.

Example 6.13. Take $f(x) = x^2$, and take $X = [-0.5, 0.5]$. For the mean value form (6.25), we generally take m to be the midpoint of X, but m must in any case lie within X. However, for the slope form, the base point y can in principle be any point in the domain of f. Let us choose $y = 1.5$. Then, the exact range $f'(X) = 2[-0.5, 0.5] = [-1, 1]$ has lower bound equal to the slope of the left dotted line in Figure 6.1 and upper bound equal to the slope of the right dotted line in Figure 6.1, while the exact range $s(f, X, y) = 1.5^2 - [-0.5, 0.5]^2 = [2, 2.25]$ has lower bound equal to the slope of the left solid line in Figure 6.1 and upper bound equal to the slope of the right solid line in Figure 6.1. The mean value form is thus

$$f(0) + [-1, 1][-0.5, 0.5] = [-0.5, 0.5],$$

while the slope form for this particular y is

$$f(1.5) + [2, 2.25]([-0.5, 0.5] - 1.5) = [-2.25, 0.25].$$

Similarly, if we take $y = -1.5$, we see by symmetry that we get an interval enclosure of $[-0.25, 2.25]$ for $f([-0.5, 0.5])$. In this case intersecting these two enclosures gives us the exact range:

$$[-2.25, 0.25] \cap [-0.25, 2.25] = [-0.25, 0.25] = f([-0.5, 0.5]).$$

Interval bounds on slopes can be computed via a process similar to that described above for computing interval bounds on partial derivatives. An early, if not the first, reference to this in the literature is [120]. Complementary to that development, in [64], Hansen proposed a technique similar to computation of slope enclosures, in which some of the variables in the expressions for interval enclosures for derivatives are replaced by points. Hansen's technique and other techniques for slope enclosures can be combined. Details, examples, and illustrations appear in [97, section 1.3.1, p. 26ff].

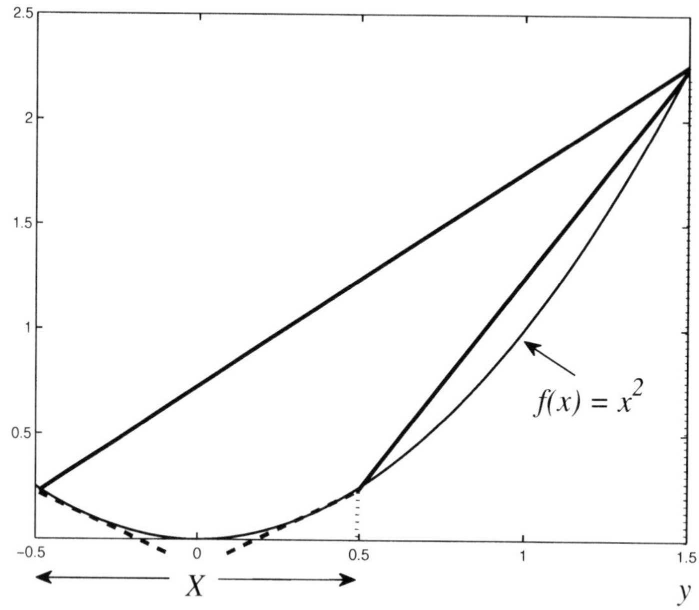

Figure 6.1. *Slope extensions versus mean value extension (Example 6.13).*

INTLAB and the Slope Form

Such slope computations are implemented in INTLAB. For information and references, type `help slopeinit` and `help slope` from the MATLAB command window. Also see [222] for an explanation of the improved version of slopes that is used in INTLAB. The following function will compute slope enclosures of the form (6.28):

```
function Fs = slope_form(f,X,y)
% Fs = slope_form(f,X,y) returns the value for the
% slope form for f evaluated over the interval X,
% with center y, as explained in Section 6.4 and
% in the references. Note that, if X and y have more
% than one coordinate, they must be represented as
% row vectors.

Xs = slopeinit(y,X);
fy = feval(f,y);
S = feval(f,Xs);
Y = X-y;
Fs = fy + S.s * Y';
```

With this function, the following MATLAB command window dialogue computes the slope form corresponding to Example 6.11:

```
>> X = [infsup(1,2), infsup(0,1)];
>> y = [1.5 0.5];
```

```
>> Fs = slope_form('example6p11',X,y)
intval Fs = [ -22.9073 , 39.6711 ]
```

Exercise 6.14. On an abstract level, slopes and centered forms are related. Explain the relationship of the centered form (6.24) to Definition 6.7 and to the slope form (6.28). (Note that people who use the term "centered form" have developed somewhat different computational techniques than the people who use the term "slope.") □

Exercise 6.15. Repeat Exercise 6.12 using the slope form. □

Exercise 6.16. In Example 6.13, we saw that different center points y in the slope or centered form (6.28) lead to different but valid enclosures and that it is not even necessary that $y \in X$. (However, y must be contained in X for the mean value form.) We also saw that the different enclosures can be intersected to obtain an even tighter enclosure; optimal choice of centers to maximize the left endpoint or minimize the right endpoint of the enclosure is studied in [20]. Redo Exercise 6.12, experimenting with different points y. □

Notes

1. From Definition 6.7, one sees that $s(f, x, y)$ is the first-order divided difference of f between x and y. Reps and Rall [216] discuss automatically computing such slopes, including higher-order divided differences.

2. More recently, Schnurr [229] has carefully described a second-order version of slopes in an interval context, as well as use of this in global optimization.

The Monotonicity Test Form

We can also obtain narrower bounds on the range of values for many functions in $\mathrm{FC}_n(X_0)$ by using a modification of the mean value form known as the *monotonicity test form*. Let us reconsider Example 6.11. Because $D_1 F(X)$ turned out to lie completely in an interval of positive real numbers, we see that $f(x_1, x_2)$ is monotonic increasing with x_1 for all (x_1, x_2) in $X = ([1, 2], [0, 1])$. Hence, we can find a lower bound on the range of values of $f(x_1, x_2)$ in X by finding a lower bound on the range of values of $f(1, x_2)$. Similarly, we can find an upper bound by finding an upper bound on the range of values of $f(2, x_2)$. For instance, we can do this with

$$f(X) \subseteq [\underline{F_{mv}(1, X_2)}, \overline{F_{mv}(2, X_2)}] \subseteq [-36.9308, 58.8966]. \tag{6.29}$$

For Example 6.11, the natural extension gives the bounds [1.7182818, 40.171074]. This is sharper than the mean value form, and is sharper on one end than the slope form given by INTLAB.

By (6.27) we have $D_2 T_1 = 2X_2$, ..., $D_2 F(X) = D_2 T_4 - D_2 T_1$. If we substitute each expression into the following ones as appropriate, we can rewrite $D_2 F(X)$ as

$$D_2 F(X) = 2X_2(e^{T_2} X_1 - 1) \tag{6.30}$$

by factoring out $2X_2$. Using (6.30), we find that

$$D_2 F([1, 2], [0, 1]) = 2[0, 1](e^{[1,2]+[0,1]^2}[1, 2] - 1) = [0, 78.342148]. \tag{6.31}$$

By (6.31), we see that $f(x_1, x_2)$ is also monotonic increasing with x_2 for all $(x_1, x_2) \in X$, so the exact range of values can be bounded sharply by computing

$$\begin{aligned} f(X) &= [F_{mv}(1, 0), F_{mv}(2, 1)] \\ &= [f(1, 0), f(2, 1)] \\ &\subseteq [2.7182818, 39.171074]. \end{aligned} \tag{6.32}$$

It follows from a theorem of Alefeld and Herzberger [13] that the mean value extension, like the centered form or slope form, has excess width of order $w(X)^2$. When $w(X)$ is small, both forms yield tight enclosures. For wider intervals, they may actually give wider enclosures than the natural extension, as in the example just discussed. However, we obtain, for the same function (6.26) with the narrower arguments $X_1 = [1, 1.01]$ and $X_2 = [.4, .401]$ from (6.25) and (6.27),

$$F_{mv}(X) = [3.029529, 3.096156],$$

whereas the natural extension

$$F(X) = X_1 e^{X_1 + X_2^2} - X_2^2$$

yields

$$F(X) = [3.029132, 3.096821].$$

The excess width of $F(X)$ is 0.001602, while that of $F_{mv}(X)$ is only 0.000539.

We can modify (6.25) to obtain the monotonicity test form $F_{mt}(X)$ as follows. Let $\mathcal{S}$ be the set of integers i (indices) such that

$$\underline{D_i F(X)} < 0 < \overline{D_i F(X)}.$$

Then, we define

$$F_{mt}(X) = [f(u), f(v)] + \sum_{i \in \mathcal{S}} D_i F(X)(X_i - m(X_i)), \tag{6.33}$$

where we specify the ith components of the point vectors u and v by

$$(u_i, v_i) = \begin{cases} (\underline{X}_i, \overline{X}_i), & \underline{D_i F(X)} \geq 0, \\ (\overline{X}_i, \underline{X}_i), & \underline{D_i F(X)} < 0 \text{ and } \overline{D_i F(X)} \leq 0, \\ (m(X_i), m(X_i)) & \text{otherwise}. \end{cases} \tag{6.34}$$

Theorem 6.3. For $f \in \mathrm{FC}_n(X_0)$ with $F(X)$ and $D_i F(X)$ $(i = 1, \ldots, n)$ expressible in the form (6.19) and extensions of its partial derivatives (e.g., (6.27)), we have $f(X) \subseteq F_{mt}(X)$ for all $X \subseteq X_0$.

Proof. Without loss of generality, let $\mathcal{S} = \{i : j \leq i \leq n\}$. Write $m_i = m(X_i)$. We have

$$f(x_1, \ldots, x_{j-1}, m_j, \ldots, m_n) \in [f(u), f(v)] \tag{6.35}$$

for all $x_i \in X_i$, $i = 1, \ldots, j-1$. Here, $u = (u_1, \ldots, u_n)$ and $v = (v_1, \ldots, v_n)$, where u_i and v_i are defined by (6.34). Now

$$f(x_1, \ldots, x_j, \ldots, x_n) - f(x_1, \ldots, m_j, \ldots, m_n) \in \sum_{i \in S} D_i F(X)(X_i - m_i) \qquad (6.36)$$

for all $x_i \in X_i$, $i = 1, \ldots, n$. Adding (6.35) and (6.36), we get $f(x) \in F_{mt}(X)$ for all $x \in X$. □

As the previous example shows, the better the extensions $D_i F(X)$ of the partial derivatives, the better the results for the form $F_{mv}(X)$. (Compare (6.32) using (6.30) with (6.29) using (6.27).)

Skelboe's Algorithm

Skelboe [232] introduced a much more efficient algorithm for computing refinements. We first seek a lower bound on the range of values of f, then repeat the process for the lower bound of $-f$ which is the upper bound of f. During the sequence of bisections, we evaluate f only on a finite subsequence of regions producing the smallest lower bounds. While the interval extensions used for the evaluations can be any of the forms discussed here (or others), there is an advantage in using forms such as the centered form or the mean value form (or the monotonicity test modification of it) because of the more rapid convergence of the excess width to zero as the size of the regions grows small.

Skelboe [232] shows that, in computing refinements, no subdivision is necessary of argument intervals for arguments which occur only once in the expression used for the interval extension. We present next a simplification of Skelboe's algorithm, known as the *Skelboe–Moore algorithm*.

Suppose we are given an interval extension $F(X)$ of $f \in \mathrm{FC}_n(X_0)$, and we want to bound the range of values of f in X_0 using refinements of F. We first find a lower bound and then, replacing $F(X)$ by $-F(X)$, repeat the procedure to be described to find an upper bound. If the evaluations are done in IA, the procedure will converge in a finite number of steps. To find a lower bound, we create a list of pairs $(Y, \underline{F(Y)})$, ordered so that $(Y, \underline{F(Y)})$ comes before $(Z, \underline{F(Z)})$ in the list only if $\underline{F(Y)} \le \underline{F(Z)}$. The interval extension used could be the mean value form, the centered form, the monotonicity test form, or any other form having inclusion isotonicity. The better the extension, the fewer the steps to convergence. The Skelboe–Moore algorithm is as follows.

Algorithm 6.1 (Skelboe–Moore). Given an interval extension F of f, a tolerance ϵ, and a box X_0, this algorithm returns a lower bound LB on the range of f over X that is within ϵ of the actual minimal value of lower bound as well as lists $\mathcal{C}$ and $\mathcal{L}$ of boxes that has been produced during the process of computing these, such that $\mathcal{C} \cup \mathcal{L} = X_0$.

(1) Set[20] $f_{\mathrm{ub}} \leftarrow f(m(X_0))$, where $f(m(X_0))$ is computed with either upward rounding or interval arithmetic[21] as $\overline{F(m(X_0))}$;

[20] We use $\leftarrow$ to denote the operation of setting a value equal to another value, while we reserve $=$ to assert that what's on the left of the $=$ is equal to what's on the right.

[21] f_{ub} represents a mathematically rigorous upper bound on the lower bound of f over X.

(2) set $X \leftarrow X_0$;

(3) initially, the lists $\mathcal{L}$ and $\mathcal{C}$ are empty;

(4) bisect X in coordinate direction i such that $w(X_i) = w(X) = \max_{1\le i\le n} w(X_i)$: $X = X^{(1)} \cup X^{(2)}$;

(5) $f_{\text{ub}} \leftarrow \min\left\{f_{\text{ub}}, \overline{F(m(X^{(1)}))}, \overline{F(m(X^{(2)}))}\right\}$;

(6) *IF* $\max\{\overline{F(X^{(1)})}, \overline{F(X^{(2)})}\} - \min\{\underline{F(X^{(1)})}, \underline{F(X^{(2)})}\} < \epsilon$, then

   (a) Place $X^{(1)}$ and $X^{(2)}$ into $\mathcal{C}$ in order

   (b) *IF* $\mathcal{L}$ is not empty *THEN*

      (i) remove the first item from $\mathcal{L}$ and place its box into X;

      (ii) *IF* $\underline{F(X)} > f_{\text{ub}}$ *THEN*

         RETURN with LB equal to lower bound on the first box in $\mathcal{C}$ and with the lists $\mathcal{C}$ and $\mathcal{L}$

      END IF

   ELSE

      RETURN with LB equal to the lower bound on the first box in $\mathcal{C}$ and with list $\mathcal{C}$

   END IF

ELSE

   (a) enter the items $(X^{(1)}, \underline{F(X^{(1)})})$ and $(X^{(2)}, \underline{F(X^{(2)})})$ in proper order in the list $\mathcal{L}$;

   (b) set $X \leftarrow$ the argument (first member of the pair) of the first item in the list $\mathcal{L}$ (with lowest $\underline{F(X)}$) and remove the item $(X, \underline{F(X)})$ from the list;

END IF

(7) *IF* $\mathcal{L}$ isn't empty *THEN* return to step (4).

Step (4) means we replace X by two interval vectors,

$$X^{(1)} = (X_1, \ldots, X_{i-1}, X_i^{(1)}, X_{i+1}, \ldots, X_n),$$
$$X^{(2)} = (X_1, \ldots, X_{i-1}, X_i^{(2)}, X_{i+1}, \ldots, X_n),$$

where

$$X_i^{(1)} = [\underline{X_i}, \tfrac{1}{2}(\underline{X_i} + \overline{X_i})], \quad X_i^{(2)} = [\tfrac{1}{2}(\underline{X_i} + \overline{X_i}), \overline{X_i}].$$

Straightforward refinement with uniform subdivision into N subintervals in each of n coordinates requires N^n evaluations of $F(X)$. On the other hand, for most functions with only a finite number of isolated extrema in X_0, the algorithm above produces bounds on the range of values, of comparable excess width, in about $Cn(\log_2 N)$ evaluations of $F(X)$ for some constant C independent of n and N.

We have implemented Algorithm 6.1 using INTLAB. Its header is as follows.

```
function [LB, X_lb, N_iter, completed_list]...
             = Skelboe_Moore (f, X0, epsilon)
% [LB, X_lb, N_iter, completed_list]...
%              = Skelboe_Moore (f, X0, epsilon)
% implements the Skelboe -- Moore algorithm in Chapter 6
% of Moore / Kearfott / Cloud.  f is the function handle
% for the objective, while X0 is the starting box, and
% epsilon is the tolerance by which the returned lower bound
% can fail to be sharp.
% On return --
%   LB is the mathematically rigorous lower bound on
%     f over X0.
%   X_lb is a box over which this lower bound has
%      been computed by interval evaluation of f.
%   N_iter is the number of bisections that were done
%      to complete the computation.
%   completed_list is the linked list of boxes produced
%      during the subdivision process.  It may be plotted
%      using "plot_list."
% It is assumed that X0 is an interval column vector.
```

To simplify programming `Skelboe_Moore`, we have written some simple linked list processing functions `new_linked_list.m`, `insert.m`, `is_empty.m`, and `remove_first.m`, as well as `bisect.m` and a plotting function `plot.m`, which uses `plotpts.m` from our GLOBSOL package.

To compare this algorithm with uniform refinement to obtain good lower and upper bounds, we have run Algorithm 6.1, using the interval extension of $f(x) = x - x^2$ as in Example 6.3, using the following MATLAB dialogue:[22]

```
>> [LB, X_lb, N_iter, completed_list] ...
     = Skelboe_Moore('example6p3',[infsup(-1,1)],1e-4)
intval LB =  [ -2.0000 , -2.0000 ]
intval X_lb = [ -1.0000 , -0.9999 ]
N_iter =    18
>> plot_list(completed_list,1,0,'Example 6.3')
>> [LB, X_lb, N_iter, completed_list]...
      = Skelboe_Moore('minus_example6p3',[infsup(-1,1)],1e-4)
intval LB =  [ -0.2501 , -0.2500 ]
intval X_lb = [ 0.5000 , 0.5001 ]
N_iter = 875
>> plot_list(completed_list,1,0,'Minus Example 6.3',1)
```

Observe that more work is required (875 bisections) to get the upper bound of 0.2500 than to get the lower bound (only 18 boxes) of -2.000 to within 10^{-4}. However, this total is much less than the comparable amount of work that would be required (about 10,000 boxes; see the table on p. 56) if a uniform subdivision were used. The reason it takes more work to obtain an upper bound that is sharp to within 10^{-4} is related to the fact that the upper bound occurs at a critical point. The basic problem is that the expression for the function

[22] Somewhat abridged. Also, note that the lower bound is output as an interval, so INTLAB will display the result rigorously rounded down.

exhibits more interval dependency due to lack of monotonicity when it is evaluated over boxes containing this critical point, so the refinement must be finer there.

In addition to the output seen in the preceding MATLAB dialogue, the routine `plot_list`, as seen in the dialogue, produces a one-dimensional graph with the intervals in the list plotted in cyclically varying colors.

Example 6.14. We will use `Skelboe_Moore.m` to find rigorous lower and upper bounds on the range of

$$f(x, y) = x^2 - 2y^2 + 5$$

over the box $X = ([-1, 1], [-1, 1])$. We programmed f in the m-files `simple_2d.m` and `minus_simple_2d.m`, and we proceed with the following dialogue:[23]

```
>> [LB, X_lb, N_iter, completed_list] ...
   = Skelboe_Moore('simple_2d',[infsup(-1,1);infsup(-1,1)],1e-4)
intval LB = [ 3.0000 , 3.0000 ]
intval X_lb = [ 0.0000 , 0.0001 ]    [ -1.0000 , -0.9999 ]
N_iter = 4417
completed_list = 1x8000 struct array
>> plot_list(completed_list,1,2,'',0)
>> [LB, X_lb, N_iter, completed_list] ...
= Skelboe_Moore('minus_simple_2d',[infsup(-1,1);infsup(-1,1)],1e-4)
intval LB = [ -6.0000 , -6.0000 ]
intval X_lb = [ 0.9999 , 1.0000 ]    [ -0.0001 , 0.0000 ]
N_iter = 856
completed_list = 1x1000 struct array
>> plot_list(completed_list,1,2,'',0)
```

This computation shows that 4417 bisections were required to get the lower bound to within 10^{-4} and 856 bisections were required to get the upper bound to within 10^{-4}. The boxes produced during the process appear in Figures 6.2 and 6.3, respectively, while a contoured surface plot of f is in Figure 6.4. (Figures 6.2 and 6.3 were produced with the calls to `plot_list` shown.) Examining the map of f in Figure 6.4, we see that the boxes are neatly clustered about the points where the lower bound and upper bound are achieved.

Since Algorithm 6.1 uses a range tolerance, and since uniform subdivision uses a domain tolerance, it is unclear how many subdivisions would be required to achieve a range tolerance of 10^{-4}. However, if it is assumed the function values vary proportionally to the increments in the variables[24] with constant of proportionality 1, then $2/10^{-4}$ subdivisions would be required in each direction, to reduce the box diameters to 10^{-4}. That is, 4×10^8 boxes would be required.

Greater efficiency can be gained by incorporating acceleration devices into the Skelboe–Moore algorithm. For example, our GlobSol software, using interval Newton methods (see Chapter 8) and other techniques, obtains the lower bound by considering just two intervals and also obtains the upper bound by considering just two intervals.[25] Furthermore, other software has compared favorably to GlobSol in the literature. We comment more on this

[23] Abbreviated.

[24] Not true at critical points.

[25] However, more work is required per interval.

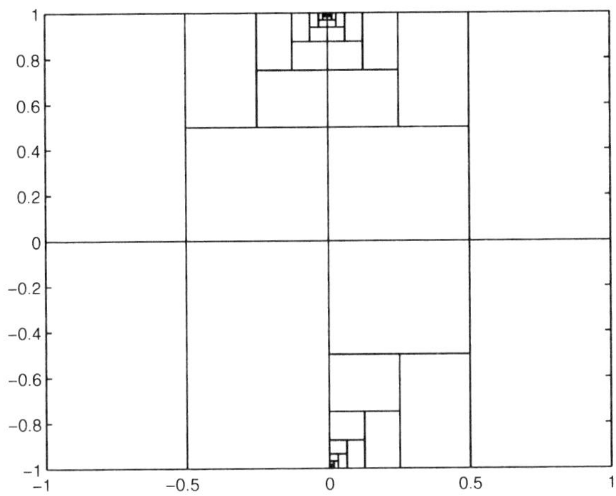

Figure 6.2. *Boxes for the lower bound in Example 6.14.*

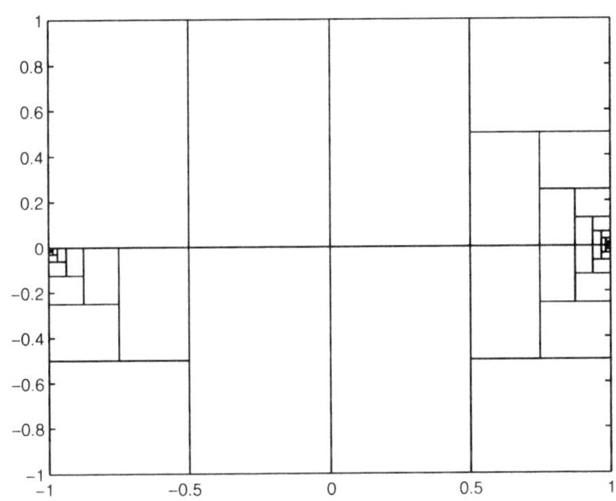

Figure 6.3. *Boxes for the upper bound in Example 6.14.*

in the next section and in Chapters 8 and 11. However, such acceleration techniques may require the function to have additional properties, such as smoothness, to work well.

Modern Context of the Skelboe–Moore Algorithm: Global Optimization

Examination of the Skelboe–Moore algorithm shows that although we have set out to find the range of f over a box X, this is equivalent to finding the *unconstrained global minimum*

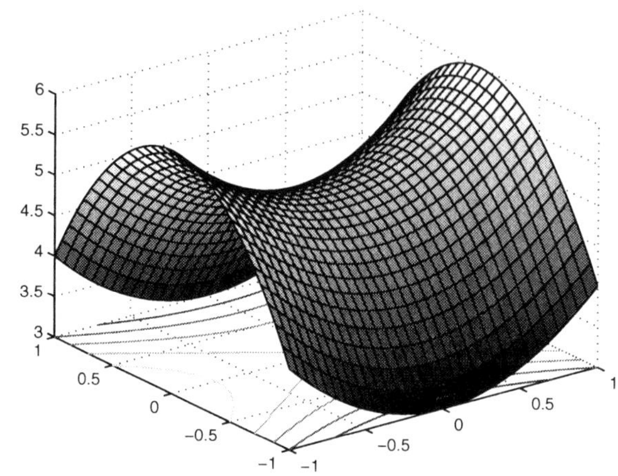

Figure 6.4. *Surface and contours for f in Example 6.14.*

of f over X. This is a difficult problem for general f but is important nonetheless. Theoretical relationships between this global optimization problem and other problems in interval computations, mathematics, and computer science are given in [121] and elsewhere.

Finding the absolute minimum of a multivariate function f over a region in $\mathbb{R}^n$ is termed *global optimization*, and f is called the *objective function*. The Skelboe–Moore algorithm is a prototypical *branch-and-bound algorithm*. Here, we "bound" the value of the objective function over a region. If the bounds are not sufficiently accurate, then we "branch," that is, we cut the region into two or more smaller regions to obtain more accurate bounds.

As a branch-and-bound algorithm, Skelboe–Moore shares much with many algorithms in commercial software for global optimization. Commercial packages for particular global optimization problems, such as *mixed integer programming* problems, use branch-and-bound methods. Furthermore, BARON (branch and reduce optimization navigator) is a general global optimization package that has won a number of prizes for software. Although BARON does not compute rigorous bounds on its answers as true interval software constructed for rigor does, BARON uses interval arithmetic in places to bound ranges. See [227, 241] for an introduction to BARON.

Interval computation researchers working in global optimization have provided many sophisticated improvements and extensions to the Skelboe–Moore algorithm while maintaining full mathematical rigor in the results. For instance, methods have been developed and improved to handle inequality and equality constraints and to speed up the process. Books and dissertations on the subject include Hansen's 1992 book [65], Hansen and Walster's 2003 update [57], [97], [213], other books, many scholarly research articles, and web resources. This is an active research area and will probably remain so for some time.

Although the Skelboe–Moore algorithm is simple and is still to be recommended in certain contexts, it is mainly of didactic and historical significance today in the context of global optimization. Algorithms in the above books, and algorithms tailored to specific applications, described in various scholarly articles, generally are more practical today. In the remainder of this section, we describe a few improvements to the Skelboe–Moore algorithm.

For functions f having only one (or perhaps a few) isolated simple maximum or minimum point(s) in X, we can use interval versions of Newton's method to further reduce the number of evaluations required to obtain sharp bounds on the range of values of f in X. See also [210].

A *clustering problem*, perhaps first described in [44] and [102, 43], is the accumulation of a large number of tiny boxes near a global minimum point of $f(x_1, \ldots, x_n)$. This phenomenon occurs, especially for large n. If interval Newton methods are used with methods for finding local optimizers, the cluster problem can be avoided with a process called *epsilon inflation*;[26] see [97, section 4.2]. Also, the cluster problem can be avoided, even for singular Hessian matrices, by astute selection of tolerances and construction of boxes about approximate optimizers. A very early example of such an algorithm is Step 4 in Algorithm 2.6 in [93]. Neumaier has further clarified the clustering problem mathematically, and gives additional insight into how to avoid it; see [170].

We will give a few more details concerning global optimization in section 11.2.

6.5 Summary

Algorithms based on refinements, discussed in this chapter, are used in interval software to reduce overestimation, often in combination with other methods we will discuss in the remaining chapters.

In the next chapter, we introduce interval computations in the context of numerical linear algebra.

[26]The first use of the term "epsilon inflation" of which we are aware occurs in [218] as "ϵ-Aufblähung," near the top of p. 99 of that work.

Chapter 7
Interval Matrices

7.1 Definitions

By an *interval matrix*, we mean a mean a matrix whose elements are interval numbers. For example, we might have

$$A = \begin{pmatrix} A_{11} & A_{12} \\ A_{21} & A_{22} \end{pmatrix} = \begin{pmatrix} [1,2] & [-1,1] \\ [0,4] & [6,8] \end{pmatrix}. \tag{7.1}$$

If A is an interval matrix with elements A_{ij} and B is a matrix with real elements B_{ij} such that $B_{ij} \in A_{ij}$ for all i and j, then we write $B \in A$.

Matrix Norm, Width, and Midpoint

We use the matrix norm

$$\|A\| = \max_i \sum_j |A_{ij}| \tag{7.2}$$

for an interval matrix A. This is an interval extension of the maximum row sum norm for real matrices. If B is any real matrix contained in an interval matrix A, then $\|B\| \le \|A\|$. We define the width $w(A)$ of an interval matrix A by

$$w(A) = \max_{i,j} w(A_{ij}). \tag{7.3}$$

The *midpoint* of A is the real matrix $m(A)$ whose elements are the midpoints of the corresponding elements of A:

$$(m(A))_{ij} = m(A_{ij}).$$

Clearly, $m(A) \in A$.

Example 7.1. For the interval matrix (7.1), we have

$$\begin{aligned} \|A\| &= \max\{|[1,2]| + |[-1,1]|\,,\ |[0,4]| + |[6,8]|\} \\ &= \max\{2+1\,,\ 4+8\} \\ &= 12, \end{aligned}$$

$$\begin{aligned} w(A) &= \max\{w([1,2]),\ w([-1,1]),\ w([0,4]),\ w([6,8])\} \\ &= \max\{1,2,4\} \\ &= 4, \end{aligned}$$

and

$$m(A) = \begin{pmatrix} m([1,2]) & m([-1,1]) \\ m([0,4]) & m([6,8]) \end{pmatrix} = \begin{pmatrix} 0.5 & 0 \\ 2 & 7 \end{pmatrix}. \qquad \square$$

Exercise 7.1. Show that the product of two interval matrices using interval arithmetic is again an interval matrix consisting of interval elements each of which is *exactly* the range of values of the corresponding element of the product of a pair of real matrices whose elements are chosen independently from the corresponding elements of the interval matrices. □

7.2 Interval Matrices and Dependency

Even though, as indicated in Exercise 7.1, the *ij*th *element* C_{ij} of the product $C = AB$ of an m by p interval matrix A and a p by n interval matrix B gives sharp bounds on the range

$$C_{ij} = \left\{ M_{ij} = \sum_{k=1}^{p} P_{ik} Q_{kj} \ :\ P_{ik} \in A_{ik} \text{ and } Q_{kj} \in B_{kj} \text{ for } 1 \le k \le p \right\}$$

for each i, $1 \le i \le m$, and each j, $1 \le j \le n$, the resulting interval matrix C may contain point matrices D that are not the result of the multiplication of point matrices $P \in A$ and $Q \in B$. The multiplication of practically any two interval matrices such that the second matrix has more than one column can serve as an example.

Example 7.2. Consider

$$A = \begin{pmatrix} [1,2] & [3,4] \end{pmatrix} \quad \text{and} \quad B = \begin{pmatrix} [5,6] & [7,8] \\ [9,10] & [11,12] \end{pmatrix}.$$

Then the product of the interval matrices A and B is

$$C = AB = \begin{pmatrix} [32,52] & [40,64] \end{pmatrix}.$$

However, $D = (32\ \ 64) \in C$, but D does not correspond to the product of any point matrix $P \in A$ with any point matrix $Q \in B$. In fact, the first element 32 of D is taken by taking the product of the lower bound matrix $\underline{A} = (1\ \ 3) \in A$ with the lower bounds $(5\ \ 9)^T$ of the first column of B, while the second element of D is obtained by taking the upper bound matrix $\overline{A} \in A$ with the upper bounds $(8\ \ 12)^T$ of the second column of B. This type of interval dependency occurs because the interval arithmetic does not assume that the same point elements are chosen from the interval elements of A in forming the sets comprising the different columns of the product interval matrix C. This is similar to interval dependency in scalar expressions, as introduced on p. 38 and treated in the Fundamental Theorem of Interval Analysis (Theorem 5.1). In terms of the Fundamental Theorem, if you think of each interval element of the left matrix A as being a variable, then this variable occurs multiple times, once for each column, in forming the columns of the interval product matrix C. □

7.3 INTLAB Support for Matrix Operations

INTLAB has extensive support for linear algebra operations involving interval matrices. INTLAB uses the MATLAB environment, which originated as a system for interactively doing numerical linear algebra and continues to have numerical linear algebra as its foundation. The following exercises deal with some elementary matrix and norm capabilities in INTLAB.

Exercise 7.2. Several functions of intervals are related to the absolute value function of a real number.

mag: mag(X), the *magnitude* of the interval X, is the real number representing the maximum absolute value of any real number in X.

mig: mig(X), the *mignitude* of the interval X, is the real number representing the minimum absolute value of any real number in X. (For example, the mignitude is used, in determining whether every matrix within an interval matrix is diagonally dominant.)

abs: abs(X) is an interval that represents the range of the absolute value function over the interval X.

INTLAB has the functions `mag`, `mig`, and `abs` that operate on intervals, interval vectors, and interval matrices. Use INTLAB to compute abs(X), mag(X), and mig(X) for the following X:

1. $X = [1, 2]$,
2. $X = [-1, 2]$,
3. $X = [-2, -1]$,
4. X the matrix A from Example 7.1.

Hint: The matrix A may be entered:

```
>> A = [infsup(1,2) infsup(-1,1)
        infsup(0,4) infsup(6,8)]
intval A =
[ 1.0000 , 2.0000 ] [ -1.0000 , 1.0000 ]
[ 0.0000 , 4.0000 ] [  6.0000 , 8.0000 ]
```

□

Exercise 7.3. INTLAB implements norm and midpoint for matrices with the functions `norm` and `mid`, respectively. The function `norm` is an extension of the MATLAB function that computes the norm of a matrix. If `A` is a real matrix, then `norm(A)` in MATLAB computes the 2-norm of `A` and is the same as `norm(A,2)`, while `norm(A,1)` and `norm(A,inf)` compute the 1-norm and ∞-norm of `A`, respectively. INTLAB does not define `norm(A,2)` when `A` is an interval matrix, but it does define `norm(A,1)` and `norm(A,inf)`; the norm defined in (7.2) corresponds to `mag(norm(A,inf))`. (The command `norm(A)` does not work when `A` is an interval matrix.) The midpoint matrix for a matrix `A` cannot be the exact midpoint matrix, but it is a floating point approximation. Use INTLAB to compute $\|A\|$ and $m(A)$ for A as in Example 7.1. □

Exercise 7.4. For scalar intervals, the INTLAB function `diam` corresponds to width. However, `diam` acts componentwise, returning a matrix when acting on a matrix.

(a) For A the matrix of Example 7.1, issue the following commands:

(i) `diam(A)`,

(ii) `max(diam(A))`,

(iii) `max(max(diam(A)))`.

Explain each result. (*Hint*: Look up `max` in MATLAB's help system.)

(b) Using this experimentation, write a MATLAB function `wid.m`, callable by `width = wid(A)`, such that the returned value is the width of the matrix `A` as defined by (7.3). □

7.4 Systems of Linear Equations

In this section, we consider finite systems of linear algebraic equations

$$Ax = b, \tag{7.4}$$

where A is an n-by-n matrix and b is an n-dimensional vector. There are two cases to consider for the coefficients of A and b:

(1) they are real numbers exactly representable by machine numbers, and

(2) they are only known to lie in certain intervals A_{ij} and B_i.

There are two types of methods for the numerical solution of such problems. *Direct methods*, such as Gaussian elimination (with or without various "pivoting" schemes) can produce exact results in case 1 in a finite number of arithmetic operations if A is nonsingular and if infinite precision arithmetic is used. *Indirect* (or *iterative*) *methods* for the solution of (7.4) in case 1 produce a sequence of approximate solutions which converge to the unique solution if, for instance, A is of the form

$$A = I - M,$$

where I is the identity matrix and M has norm less than 1, such as for the norm (maximum row sum norm)

$$\|M\| = \max_i \sum_j |M_{ij}|. \tag{7.5}$$

An exact solution in case (2) is much more difficult; the set of all solutions to (7.4), when the coefficients of A and b can range over intervals, may be a very complicated set,[27] and its computation is, in general, an NP-hard problem [121]. An example of such a solution set, with

$$A = \begin{pmatrix} [\ \ 2,4] & [-2,1] \\ [-1,2] & [\ \ 2,4] \end{pmatrix}, \qquad B = \begin{pmatrix} [-2,2] \\ [-2,2] \end{pmatrix}, \tag{7.6}$$

[27] See, for instance [97, section 1.2.1] and the references there for a discussion of different kinds of solution sets.

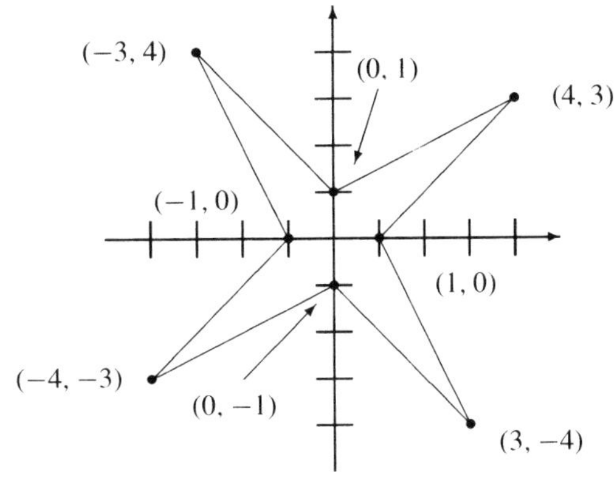

Figure 7.1. *Exact solution set to $Ax = B$, A, and B in Equation (7.6).*

appears in Figure 7.1 (from [97, p. 20]). However, such a solution set can be enclosed in an interval vector (an n-dimensional rectangle). As an immediate consequence of Corollary 5.1, we have the following computable test for existence of solutions to (7.4), with real or interval coefficients.

Theorem 7.1. If we can carry out all the steps of a direct method for solving (7.4) in IA (if no attempted division by an interval containing zero occurs, nor any overflow or underflow), then the system (7.4) has a unique solution for every real matrix in A and every real vector in b, and the solution is contained in the resulting interval vector X.

Consider the general 2-by-2 linear system

$$a_{11}x_1 + a_{12}x_2 = b_1,$$
$$a_{21}x_1 + a_{22}x_2 = b_2.$$

Formally, we can eliminate x_1 from the second equation by subtracting term by term a_{21}/a_{11} times the first equation, as done in Gaussian elimination. Doing that, we can solve for x_2 by a division from the equation

$$(a_{22} - (a_{21}/a_{11})a_{12})x_2 = b_2 - (a_{21}/a_{11})b_1.$$

Then we can use that value of x_2 and the first equation to solve for x_1 by another division from

$$a_{11}x_1 = b_1 - a_{12}x_2.$$

We can carry out all that numerically with interval coefficients for all the a's and b's to obtain intervals X_1 and X_2 containing all the possible solutions (x_1, x_2) for any choices of real coefficients in the corresponding interval coefficients. All this can be done *if* the interval for a_{11} does not contain zero *and if* the computed interval containing $a_{22} - (a_{21}/a_{11})a_{12}$ does not contain zero.

The determinant of the matrix is the product of those two quantities: $a_{11}(a_{22} - (a_{21}/a_{11})a_{12})$. The conditions stated above are sufficient for the matrix to be nonsingular for *any* choice of real coefficients in the interval coefficients.

Exercise 7.5. The mesh equations for an electric circuit are expressed as

$$\begin{pmatrix} R_1 + R_2 & -R_2 \\ -R_2 & R_2 + R_3 \end{pmatrix} \begin{pmatrix} I_1 \\ I_2 \end{pmatrix} = \begin{pmatrix} V_1 \\ -V_2 \end{pmatrix}.$$

Take $V_1 = 10$, $V_2 = 5$, and $R_1 = R_2 = R_3 = 1000 \pm 10\%$. Find enclosures for I_1 and I_2. □

In case (1), with exactly representable real coefficients, if the hypothesis of Theorem 7.1 is satisfied, the resulting interval vector has a width which approaches zero as the number of digits carried (in the IA) increases.

Example 7.3. Consider the following ill-conditioned system:

$$\begin{aligned} 2.0000x_1 + 3.001x_2 &= 1.0000, \\ 0.6667x_1 + 1.000x_2 &= 0.3333. \end{aligned}$$

We will carry out a direct method for solving the system using n-place IA for $n = 4, 5, 6, 7, 8$, and 9. We can eliminate the x_1 term in the second equation to obtain

$$(1.000 - (0.6667/2.0000)(3.001))x_2 = 0.3333 - (0.6667/2.0000)(1.0000).$$

If we carry out the arithmetic using n-place IA, we obtain (for x_2):

for $n = 4$,

$$\begin{aligned} &0.6667/2.0000 \in [0.3333, 0.3334], \\ &(0.6667/2.0000)(3.001) \in [1.000, 1.001], \\ &.3333 - (.6667/2.000)(1.000) \in [-.0001, 0], \\ &(1.000 - (0.6667/2.0000)(3.001)) \in [-0.0010, 0], \\ &x_2 \in [0, \infty) \quad \text{(no upper bound on } x_2\text{);} \end{aligned}$$

for $n = 5$,

$$\begin{aligned} &0.6667/2.000 \in [.33335, .33335], \\ &(0.6667/2.000)(3.001) \in [1.0003, 1.0004], \\ &0.3333 - (0.6667/2.000)(1.000) \in [-0.00005, -0.00005], \\ &(1.000 - (0.6667/2.000)(3.001)) \in [-0.00040, -0.00030], \\ &x_2 \in [0.\underline{1}2500, 0.\underline{1}6667]; \end{aligned}$$

for $n = 6$, $x_2 \in [0.\underline{1}28205, 0.\underline{1}31579]$;

for $n = 7$, $x_2 \in [0.\underline{130}2083, 0.\underline{130}5484]$;

for $n = 8$, $x_2 \in [0.\underline{1304}1210, 0.\underline{1304}4613]$;

for $n = 9$, $x_2 \in [0.\underline{130429111}, 0.\underline{130429112}]$.

The sudden increase in accuracy in this simple example which occurs between $n = 8$ and $n = 9$ is accounted for by the fact that, beyond $n = 8$, the only roundoff error remaining is in the final division for x_2. In this example, the width of the computed interval goes down by at least 10^{-1} for each increase of a unit in n beyond $n = 5$. □

A phenomenon, somewhat puzzling at first, was observed during early experiments with interval versions of Gaussian elimination. In case (1), with real coefficients, the midpoint of the resulting interval solution in IA is usually very much closer to the exact solution than the width of the interval solution—by seven or eight places for systems of order around 20. For a fixed number of digits carried in the IA operations, the width of the resulting interval vector increases as the order n increases for most matrices at about the same rate as the worst-case analysis of von Neumann and Goldstine [246]. During the Gaussian elimination process, there are multiple occurrences of the coefficients in combination with subtractions and divisions which reverse endpoints, resulting in the observed excess width (in most cases) of the interval results. However, for special classes of matrices, interval Gaussian elimination can give good results. Interval Gaussian elimination can always be carried out, even without pivoting, when A is strongly (strictly) diagonally dominant, that is, when

$$|A_{ii}| > \sum_{j \neq i} |A_{ij}| \qquad (i = 1, \dots, n).$$

See [13] and [57] for further discussion. Also see the section A Common Misuse of Interval Arithmetic in the demo for INTLAB.

First Look at the Krawczyk Method

Iterative methods for interval linear systems were begun by Hansen, e.g., [57]. One such method is a linear version of the *Krawczyk method*, which we cover more completely in section 8.2. We can multiply both sides of (7.4) by a matrix Y (for instance, an approximate inverse of $m(A)$) and define

$$E = I - YA.$$

If $\|E\| < 1$ using (7.2), then the sequence

$$X^{(k+1)} = \{Yb + EX^{(k)}\} \cap X^{(k)} \qquad (k = 0, 1, 2, \dots) \tag{7.7}$$

with

$$X_i^{(0)} = [-1, 1] \, \|Yb\| \, / (1 - \|E\|) \qquad (i = 1, \dots, n)$$

is a nested sequence of interval vectors containing the unique solution to (7.4) for every real matrix in A and every real vector in b. In IA, the sequence (7.7) converges in a finite number of steps to an interval vector containing the set of solutions to (7.4). Thus, we have the next theorem.

Theorem 7.2. Using the norm (7.2), the system (7.4) has a unique solution x for every real matrix in A and every real vector in b (for interval matrix A and interval vector b) if $\|I - YA\| < 1$ for some matrix Y (real or interval). Furthermore, the solution vector x is contained in the interval vector $X^{(k)}$ defined by (7.7) for every $k = 0, 1, \dots$. Using IA, the sequence $\{X^{(k)}\}$ converges in a finite number of steps to an interval vector containing the set of solutions to (7.4).

Example 7.4. Consider (7.4) with

$$A = \begin{pmatrix} 3 & 1 \\ 3 & 2 \end{pmatrix}, \quad b = \begin{pmatrix} 1 \\ 0 \end{pmatrix}.$$

If we choose

$$Y = \begin{pmatrix} 0.6 & -0.3 \\ -1 & 1 \end{pmatrix}, \quad \text{an approximate inverse of } A,$$

we find, using three-digit IA,

$$E = \begin{pmatrix} 0.1 & 0 \\ 0 & 0 \end{pmatrix}, \quad Yb = \begin{pmatrix} 0.6 \\ -1 \end{pmatrix}, \quad X^{(0)} = \begin{pmatrix} [-1.12, 1.12] \\ [-1.12, 1.12] \end{pmatrix},$$

and we obtain

$$X^{(1)} = \begin{pmatrix} [0.478, 0.712] \\ -1 \end{pmatrix}, \quad X^{(2)} = \begin{pmatrix} [0.647, 0.672] \\ -1 \end{pmatrix},$$

$$X^{(3)} = \begin{pmatrix} [0.664, 0.668] \\ -1 \end{pmatrix}, \quad X^{(4)} = \begin{pmatrix} [0.666, 0.667] \\ -1 \end{pmatrix},$$

with $X^{(k)} = X^{(4)}$ for $k \geq 4$. □

The iterative method (7.7) is also applicable in case A and b have interval components. Alternate methods and methods that give sharper bounds on the set of solutions to linear systems with interval coefficients may be found in [13], [57], [218], and [221]. Method (7.7) is a linear version of the Krawczyk method that we will explain in more detail in section 8.2. In fact, because nonlinearities in a system can be encompassed by intervals in a linear system, as we shall see in the next chapter, in some contexts there is less distinction between interval methods for linear systems and interval methods for nonlinear systems than there is with point methods.

Additional Notes

1. We will say more about convergence of the Krawczyk method in section 8.2.
2. Better enclosures of the solution can be obtained in practice through a defect-correction scheme than with just the simple method we have presented here. This was observed in [218, section 2.b, p. 53], was later explained in [221], and is implemented within routine `verifylss` in INTLAB.

7.5 Linear Systems with Inexact Data

There are a few special types of matrices for which satisfactory results can be obtained. Unfortunately, the use of direct methods, such as Gaussian elimination with interval arithmetic, cannot be recommended in the general case for finding enclosures of the set of all solutions to linear systems with inexact data, represented by interval coefficients. Bounds of intermediate quantities tend to grow rapidly especially because of dependence among generated intervals and the endpoint reversals occurring in subtractions. The exact solution set can be quite complicated, but we can nevertheless compute an enclosure of it, using the iterative method outlined in the previous section. Let us revisit Exercise 7.5 to illustrate its application.

Example 7.5. The mesh equations for an electric circuit are expressed as

$$\begin{pmatrix} R_1 + R_2 & -R_2 \\ -R_2 & R_2 + R_3 \end{pmatrix} \begin{pmatrix} I_1 \\ I_2 \end{pmatrix} = \begin{pmatrix} V_1 \\ -V_2 \end{pmatrix}$$

with $V_1 = 10$, $V_2 = 5$, and $R_1 = R_2 = R_3 = 1000 \pm 10\%$. We can use the iterative method (7.7) to find enclosures for I_1 and I_2. In this example, the matrix A in (7.7) can be written

$$A = \begin{pmatrix} [1800, 2200] & -[900, 1100] \\ -[900, 1100] & [1800, 2200] \end{pmatrix},$$

and the right-hand side vector is

$$b = \begin{pmatrix} 10 \\ -5 \end{pmatrix}.$$

Now the midpoint matrix is

$$m(A) = \begin{pmatrix} 2000 & -1000 \\ -1000 & 2000 \end{pmatrix} = 2000 \begin{pmatrix} 1 & -\frac{1}{2} \\ -\frac{1}{2} & 1 \end{pmatrix},$$

which has the inverse

$$[m(A)]^{-1} = \frac{1}{2000} \begin{pmatrix} \frac{4}{3} & \frac{2}{3} \\ \frac{2}{3} & \frac{4}{3} \end{pmatrix} = \begin{pmatrix} \frac{4}{6000} & \frac{2}{6000} \\ \frac{2}{6000} & \frac{4}{6000} \end{pmatrix}$$

$$= \begin{pmatrix} 6.666\ldots \times 10^{-4} & 3.333\ldots \times 10^{-4} \\ 3.333\ldots \times 10^{-4} & 6.666\ldots \times 10^{-4} \end{pmatrix}.$$

It is not necessary to find the exact inverse, although in this simple case, we can do it. In fact, it will suffice to choose as our approximation to $[m(A)]^{-1}$ the matrix

$$Y = 10^{-4} \begin{pmatrix} 6 & 3 \\ 3 & 6 \end{pmatrix}.$$

What we need is a matrix Y that makes the norm of $E = I - YA$ smaller than 1, and the smaller the better. Here, we have

$$E = \begin{pmatrix} 1 & 0 \\ 0 & 1 \end{pmatrix} - \begin{pmatrix} 6 & 3 \\ 3 & 6 \end{pmatrix} \begin{pmatrix} [0.1800, 0.2200] & -[0.0900, 0.1100] \\ -[0.0900, 0.1100] & [0.1800, 0.2200] \end{pmatrix}$$

$$= \begin{pmatrix} [-0.05, 0.25] & [-0.12, 0.12] \\ [-0.12, 0.12] & [-0.05, 0.25] \end{pmatrix},$$

so $\|E\| = 0.37 < 1$. It follows that we can obtain a nested sequence of enclosures of the set of solutions to the linear system with inexact data (known only to be in the given intervals), from the iterative method (7.7), beginning with

$$X^{(0)} = \frac{\|Yb\|}{1 - \|E\|} \begin{pmatrix} [-1, 1] \\ [-1, 1] \end{pmatrix},$$

where

$$Yb = 10^{-4}\begin{pmatrix} 6 & 3 \\ 3 & 6 \end{pmatrix}\begin{pmatrix} 10 \\ -5 \end{pmatrix} = \begin{pmatrix} 0.0045 \\ 0 \end{pmatrix},$$

so, with outward rounding to two digits, we can choose

$$X^{(0)} = \begin{pmatrix} [-0.0072, 0.0072] \\ [-0.0072, 0.0072] \end{pmatrix}.$$

Thus we already know at this point that

$$-7.2 \times 10^{-3} \leq I_1 \leq 7.2 \times 10^{-3},$$
$$-7.2 \times 10^{-3} \leq I_2 \leq 7.2 \times 10^{-3}$$

for every solution of the mesh equations for the given electric circuit, within the uncertainty specified for the inexact data. We can try to improve this initial enclosure, using the interval iteration formula (7.7), which for this example is

$$\begin{pmatrix} X_1^{(k+1)} \\ X_2^{(k+1)} \end{pmatrix} = \left\{ \begin{pmatrix} 0.0045 \\ 0 \end{pmatrix} + \begin{pmatrix} [-0.05, 0.25] & [-0.12, 0.12] \\ [-0.12, 0.12] & [-0.05, 0.25] \end{pmatrix} \begin{pmatrix} X_1^{(k)} \\ X_2^{(k)} \end{pmatrix} \right\} \cap \begin{pmatrix} X_1^{(k)} \\ X_2^{(k)} \end{pmatrix}.$$

From

$$X^{(0)} = \begin{pmatrix} X_1^{(0)} \\ X_2^{(0)} \end{pmatrix} = \begin{pmatrix} [-0.0072, 0.0072] \\ [-0.0072, 0.0072] \end{pmatrix},$$

we obtain

$$\begin{pmatrix} 0.0045 \\ 0 \end{pmatrix} + \begin{pmatrix} [-0.05, 0.25] & [-0.12, 0.12] \\ [-0.12, 0.12] & [-0.05, 0.25] \end{pmatrix} \begin{pmatrix} [-0.0072, 0.0072] \\ [-0.0072, 0.0072] \end{pmatrix}$$
$$= \begin{pmatrix} [0.001836, 0.0117] \\ [-0.002664, 0.002664] \end{pmatrix}.$$

Hence, the set of solutions also is contained in

$$\begin{pmatrix} X_1^{(1)} \\ X_2^{(1)} \end{pmatrix} = \begin{pmatrix} [0.001836, 0.0117] \\ [-0.002664, 0.002664] \end{pmatrix} \cap \begin{pmatrix} [-0.0072, 0.0072] \\ [-0.0072, 0.0072] \end{pmatrix}$$
$$= \begin{pmatrix} [0.001836, 0.0072] \\ [-0.002664, 0.002664] \end{pmatrix}.$$

Another iteration yields

$$\begin{pmatrix} 0.0045 \\ 0 \end{pmatrix} + \begin{pmatrix} [-0.05, 0.25] & [-0.12, 0.12] \\ [-0.12, 0.12] & [-0.05, 0.25] \end{pmatrix} \begin{pmatrix} [0.001836, 0.0072] \\ [-0.002664, 0.002664] \end{pmatrix}$$
$$= \begin{pmatrix} [0.00382032, 0.00661968] \\ [-0.00153, 0.00153] \end{pmatrix},$$

so the set of solutions is also contained in

$$\begin{pmatrix} X_1^{(2)} \\ X_2^{(2)} \end{pmatrix} = \begin{pmatrix} [0.00382032, 0.00661968] \\ [-0.00153, 0.00153] \end{pmatrix} \cap \begin{pmatrix} [0.001836, 0.0117] \\ [-0.002664, 0.002664] \end{pmatrix}$$

$$= \begin{pmatrix} [0.00382032, 0.00661968] \\ [-0.00153, 0.00153] \end{pmatrix}.$$

We could continue the iterations to further sharpen the enclosure. The computations displayed above illustrate the interval operations needed.

For this example, we can compare the enclosure above with the one found in the solution of Exercise 7.5. Using a direct method for this small linear system, we found the sharper enclosure

$$I_1 = [0.00433, 0.00582], \quad I_2 = [-0.000419, 0.000419]. \qquad \square$$

Exercise 7.6. Use INTLAB to check the above computations. *Hint*: There are several ways of doing this. One is to check the computations directly by entering the matrices and vectors, then doing the matrix-vector multiplications, taking midpoints, inversions, etc., in INTLAB. You can also use the function `Krawczyk_step` displayed on p. 118 with $y = 0$ and $f(y) = Ay - b$, but you will need to check the starting vector separately. Here is a set of commands to get you started:

```
A = [infsup(1800,2200), -infsup(900,1100);...
     -infsup(900,1100), infsup(1800,2200)]
B = [10;-5]
Y = 10^(-4) * [6 3; 3 6]
E = eye(2) - Y*A
xmult = norm(Y*B)/(1-mag(norm(E,inf)))
X = xmult * [infsup(-1,1);infsup(-1,1)]
```

(Above, you see `mag(norm(E,inf))` because INTLAB computes the matrix norm as an interval, and `mag(norm(E,inf))` corresponds to the matrix norm we have defined.) You can then duplicate the first step of the above computations via

```
X1 = Y*B + E*X
```

etc. Alternately, to use `Krawczyk_step`, issue the line

```
X = intersect(Krawcayk_step(X,[0.0],'example7p5'),X)
```

where `example7p5` (also supplied) is as follows:

```
function [val] = example7p5(x)
% [y] = example7p5(x) implements Example 7.5 for
% Krawczyk_step
A = [ infsup(1800,2200) -infsup(900,1100)
     -infsup(900,1100)   infsup(1800,2200)];
b = [midrad(10,0);midrad(-5,0)];
val = A*x - b;
```

You can then iterate the Krawczyk method by pressing the up-arrow key followed by the return key. However, the results from `Krawczyk_step` will differ a bit from those on the preceding pages, because `Krawczyk_step` uses a more accurate approximation to $m(A)^{-1}$ for Y. (See p. 200 for other possible discrepancies.) $\square$

Exercise 7.7. The INTLAB routine `verifylss` computes verified enclosures to linear systems of equations using a combination of the Krawczyk method and an improved method originating from work of Hansen, [218], and [184], [168]; type `help verifylss` from the MATLAB command window for more information. Use `verifylss` on the linear system just worked out in Example 7.5 and compare the results you obtain with the above results. □

An Additional Note As mentioned, we will treat the Krawczyk method more generally in section 8.2, starting on p. 116. There, we treat it as a method for bounding the solutions to nonlinear systems as well as give additional background.

Interval Gauss–Seidel

An alternate method, also commonly used in practice, was originally introduced by Hansen et al. in the Hansen–Sengupta method [66] and is an interval version of classical Gauss–Seidel iteration. As in the Krawczyk method, assume A is an interval matrix, B is an interval vector (where either A or B can have zero width or nonzero width), and we wish to bound the solution set to

$$Ax = B. \tag{7.8}$$

We use a preconditioner matrix Y typically but not necessarily chosen to be $m(A)^{-1}$ (see [97, Chapter 3]), obtaining

$$Gx = C, \quad \text{where } G = YA \text{ and } C = YB. \tag{7.9}$$

Formally solving the ith row of the preconditioned system (7.9) for x_i, and assuming we iterate to obtain a new range for X_i by substituting the ranges for X_j, $j \neq i$, using the most recent value of X_j wherever possible, we obtain the iteration equations for the interval Gauss–Seidel method:

$$\begin{aligned}
X_i^{(0)} \quad & \text{is given,} && 1 \le i \le n,\\
\tilde{X}_i^{(k+1)} \quad & \leftarrow \quad \frac{1}{G_{i,i}}\left\{C_i - \sum_{j=1}^{i-1} G_{i,j}X_j^{(k+1)} - \sum_{j=i+1}^{n} G_{i,j}X_j^{(k)}\right\}, && i = 1, 2, \ldots, n,\\
X_i^{(k+1)} \quad & \leftarrow \quad \tilde{X}_i^{(k+1)} \cap X_i^{(k)},
\end{aligned} \tag{7.10}$$

where $G = \{G_{i,j}\}$.

In the Krawczyk method, as long as a preconditioner matrix Y is available, the iteration (7.7) contains no divisions and consists of a single box. However, it could happen that $0 \in G_{i,i}$, where the diagonal element $G_{i,i}$ of G is the denominator in (7.10); thus, analogous to what we will see in section 8.1 for the univariate interval Newton method, extended interval arithmetic can be used, and the result of a step of (7.10) can be two boxes. The following m-file computes a step of (7.10) for a single i without intersecting. (That is, `gauss_seidel_step` evaluates the right side of the middle assignment in (7.10) for a given i.) The preconditioner is not given as a matrix but rather is given as a row Y_i to allow the preconditioner matrix to be computed rowwise, as explained in [97, Chapter 3]. The result of a division by zero is computed according to the Cset theory of [194], using the variant that views $-\infty$ and ∞ as part of the number system. (We explain Cset theory on p. 113.)

```
function [new_X_i, second_new_X_i, there_are_2,...
    numerator, denominator]...
    = gauss_seidel_step(i, Y_i, A, B, X)
% [new_X_i, second_new_X_i, there_are_2]...
%  = gauss_seidel_step(i, Y_i, A, B, X) returns
% the result of applying a Gauss--Seidel step with variable i,
% preconditioner matriX Y_i, and initial guess X.  The variable
% there_are_2 is set to 1 if 2 semi-infinite intervals are returned,
% in which case second_new_X_i has the second interval; there_are_2
% is set to 0 and second_new_X_i is not set if there is only
% one interval returned.

   n = size(A,2);
   G_i = Y_i*A;
   C_i = Y_i*B;
   numerator = C_i;
   new_X_i = X(i);
   second_new_X_i = X(i);
   if (n > 1)
       if (i > 1)
           numerator = numerator - G_i(1:i-1)*X(1:i-1);
       end
       if (i < n)
          numerator = numerator - G_i(i+1:n)*X(i+1:n);
       end
   end
   denominator = G_i(i);
   numerator;
   denominator;
   if (~in(0,denominator))
       there_are_2 = 0;
       new_X_i = numerator / denominator;
   elseif (~in(0,numerator))
       there_are_2 = 1;
       supnum = sup(numerator);
       if(supnum < 0)
           if sup(denominator)==0
               tmp1 = infsup(-Inf,-Inf);
           else
               tmp1 = midrad(supnum,0) / midrad(sup(denominator),0);
           end
           if inf(denominator) == 0
               tmp2 = infsup(Inf,Inf);
           else
               tmp2 = midrad(supnum,0) / midrad(inf(denominator),0);
           end
```

```
            new_X_i = infsup(-Inf,sup(tmp1));
            second_new_X_i = infsup(inf(tmp2),inf);
        else
            infnum = inf(numerator);
            if inf(denominator)==0
                tmp1 = infsup(-Inf,-Inf)
            else
                tmp1 = midrad(infnum,0) / midrad(inf(denominator),0);
            end
            if sup(denominator) == 0
                tmp2 = infsup(Inf,Inf)
            else
                tmp2 = midrad(infnum,0) / midrad(sup(denominator),0);
            end
            new_X_i = infsup(-Inf,sup(tmp1));
            second_new_X_i = infsup(inf(tmp2),Inf);
        end
    else
        there_are_2=0;
        new_X_i = infsup(-Inf,Inf);
    end
end
```

The following m-file repeatedly calls gauss_seidel_step to do an entire sweep of (7.10), that is, to compute $X_i^{(k+1)}$, $1 \le i \le n$, given $X_i^{(k)}$, $1 \le i \le n$:

```
function [X_kp1,is_empty, error_occurred] =...
  Gauss_Seidel_image(A, B, X_k)
% X_kp1] = Gauss_Seidel_image(A,B,X_k) returns the image after a
% sweep of Gauss--Seidel iteration ( that is, (7.8) of the text)
% for the interval linear system A X = B, beginning with box X_k,
% 1 <= i <= n.
% This is done using the inverse midpoint preconditioner.
% Upon return:
% if error_occurred = 1, then the computation could not proceed.
% (For example, the midpoint preconditioner may have been
% singular, or the denominator may have contained zero; the
% case of more than one box in the image is not handled
% with this routine.) Otherwise, error_occurred = 0.
% If error_occurred = 0 but is_empty = 1, this means that
% an intersection of a coordinate extent was empty.  In this
% case, there are no solutions to A X = B within X_k.
% If error_occurred = 0 and is_empty = 0, then the image under
% the Gauss--Seidel sweep is returned in X_kp1.

n = length(B);
Y = inv(mid(A));

error_occurred = 0;
is_empty = 0;
i=1;
```

```
X_kp1 = midrad(zeros(n,1),0);
while(~error_occurred & ~is_empty & i<= n)
    [new_x_i, second_new_x_i, there_are_2, num, denom] ...
        = gauss_seidel_step(i, Y(i,:), A, B, X_k);
    if(there_are_2)
        error_occurred = 1;
    end
    if (~error_occurred)
        is_empty = isempty_(intersect(new_x_i,X_k(i)));
        if (~is_empty)
            X_kp1(i) = intersect(new_x_i, X_k(i));
        end
    end
    i=i+1;
end
```

We may use `Gauss_Seidel_image` to bound the solution set in Example 7.5 as follows:

```
>> A = [ infsup(1800,2200) -infsup(900,1100)
-infsup(900,1100)    infsup(1800,2200)]
intval A =
  1.0e+003 *
[ 1.8000 ,   2.2000 ] [ -1.1000 , -0.9000 ]
[-1.1000 , -0.9000 ] [   1.8000 , 2.2000 ]
>> b = [midrad(10,0);midrad(-5,0)]
intval b =
[ 10.0000 , 10.0000 ]
[ -5.0000 , -5.0000 ]
>> X0 = [infsup(-1,1), infsup(-1,1)]
intval X0 =
[ -1.0000 , 1.0000 ] [ -1.0000 , 1.0000 ]
>> [X1,is_empty, error_occurred] = Gauss_Seidel_image(A, b, X0)
intval X1 =
[ -0.1541 , 0.1661 ]
[ -0.1601 , 0.1601 ]
is_empty = 0
error_occurred = 0
  .
  .
  .
>> [X6,is_empty, error_occurred] = Gauss_Seidel_image(A, b, X5)
intval X6 =
[ 0.0041  , 0.0062 ]
[ -0.0011 , 0.0011 ]
is_empty = 0
error_occurred = 0
>> [X7,is_empty, error_occurred] = Gauss_Seidel_image(A, b, X6)
intval X7 =
[ 0.0041 , 0.0062 ]
[ -0.0010 , 0.0010 ]
```

It is shown in [167] that for a given preconditioner (e.g., the inverse midpoint matrix), the interval Gauss–Seidel method gives an ultimate result at least as narrow as the Krawczyk method. Furthermore, the interval Gauss–Seidel method can be used effectively in cases when the solution set is unbounded and for nonlinear systems; see [66] for the first such use, and see [97] for more development, including use of preconditioners other than the inverse midpoint matrix. For advanced theory comparing various methods, see [167].

We also can solve the linear system for selected real data from within the given intervals of uncertainty. In fact, a widely used approach to estimating the effects of uncertainties in data upon solutions to equations is the Monte Carlo method, in which one uses random number generators to sample the ranges or distributions of input data, solves the problem for each sample input vector, then summarizes the results. Often, in practice, this leads to solving the equations millions of times to get estimated output distributions. The interval approach is cruder, but faster.

Further discussion of interval methods for enclosing the set of solutions to linear systems with interval coefficients may be found in [13], [57] and in various other references.

An Additional Note The interval Gauss–Seidel method as illustrated in our example routine `Gauss_Seidel_image` could be made to execute more quickly with faster implementations of interval arithmetic. Also, designers of methods for verification of solutions generally follow a principle of using floating point arithmetic to first obtain good approximations, wherever possible, resorting to interval computations only where necessary. From this point of view,[28] the interval Gauss–Seidel method requires more interval arithmetic than certain other methods of enclosing solution sets.

7.6 More on Gaussian Elimination

Gaussian elimination is an example of an algorithm that cannot be converted directly into an interval-arithmetic-based algorithm by simply replacing floating point operations by interval operations. This is true even if partial pivoting is used.

It is well-known that, in the worst case, the errors in the elements of the triangular factors in Gaussian elimination carried out in floating point arithmetic can grow proportionally to 2^{n-1}, where n is the number of equations and unknowns. This is shown with the famous example of Wilkinson [248, p. 212]. However, Gaussian elimination with partial pivoting, carried out in floating point arithmetic, does not exhibit undue magnification of roundoff error for most matrices that occur in practice, although Stephen Wright has presented a class of matrices that are important in practice, for which Gaussian elimination with partial pivoting is unstable [249].

In contrast, as we explained on p. 91, if we begin with a point matrix A and point right-hand-side vector b, but then use interval arithmetic in the Gaussian elimination algorithm to obtain verified bounds on the solution vector x, the typical result is that the widths of the resulting bounds X will be on the order of 2^{n-1} times the rounding error unit, except for certain special matrices.[29] Furthermore, there is more of a question about how the pivot

[28]Assuming that interval computations are expensive compared to ordinary floating point computations; however, fast implementations in software can be close to a factor of 5.

[29]Such as diagonally dominant matrices, M-matrices, or H-matrices.

element should be chosen, since the pivots become intervals.[30] Indeed, it is likely that all potential pivots in the jth column will contain zero at some step, due to magnification of widths, so that production of finite bounds on the solution becomes impossible.

There are various ways in which Gaussian elimination can be used to get tight bounds on the solutions to point systems of linear equations. For instance, one may precondition, just as we did for the interval Gauss–Seidel method or, in a way, with the matrix Y in the Krawczyk method. That is, we first use floating point arithmetic to produce an approximate inverse $Y \approx A^{-1}$, then use interval arithmetic to do the matrix-matrix multiplication and matrix-vector multiplication, to produce an interval matrix G that contains the exact matrix YA and an interval vector C that contains the exact vector Yb. The solution set to the system of equations $Gx = C$ must then contain the exact solution to the original noninterval system $Ax = b$. The Gaussian elimination algorithm can then be performed on $Gx = C$, leading to reasonable bounds on the solution set to $Ax = b$.

If the original system $Ax = B$ is an interval system of equations, that is, if the elements of the matrix A and the vector B have nonzero widths at the start, we may multiply by a preconditioner matrix $Y \approx m(A)^{-1}$, just as we did for the interval Gauss–Seidel method. The Gaussian elimination algorithm is then more likely to succeed for the resulting system of equations.

Interval Gaussian elimination is analyzed theoretically in [12], [13], and [167] as well as in other papers and reference works.

7.7 Sparse Linear Systems Within INTLAB

MATLAB has a sparse matrix format, and the routines in INTLAB (and in particular `verifylss`) can use this sparse format. Depending upon how wide the intervals are, the structure of the matrix, and the condition of the matrix, meaningful answers can be obtained for some very large matrices. Here, we will give an example with cursory explanation; complete understanding of this example can be obtained by studying the sparse matrix format within MATLAB's help system.

Example 7.6. Suppose we wish to bound the solutions to $Ax = b$, where A is the 1000 by 1000 matrix given by

$$A = \begin{pmatrix} [2.9999, 3.0001] & -1 & 0 & \cdots & \\ -1 & [2.9999, 3.0001] & -1 & 0 & \cdots \\ & & & & \\ \vdots & & \ddots & & \\ & & & & \\ 0 & \cdots & 0 & -1 & [2.9999, 3.0001] \end{pmatrix},$$

and $b \in \mathbb{R}^{1000}$ has all entries equal to -1. If such a matrix is passed to `verifylss`, there is an implicit assumption that the diagonal entries vary independently, that is, that the diagonal matrices A are x_i where, in general $x_i \neq x_j$, although each x_i lies in the interval

[30] One way of choosing a pivot element during the jth step would be to choose $A_{i_0,j}$ to be the pivot row, where $i_0 = \text{argmax}_i \{\text{mig}(A_{i,j})\}$, where "mig" is the mignitude defined on p. 87.

[2.9999, 3.0001]. (If the diagonal entries are all the same value, even though that value is known to lie only within [2.9999, 3.0001], then sharper bounds on the solution set possibly can be obtained by taking account of this fact.)

A MATLAB dialogue that gives us bounds on the solution set follows:

```
>> n = 1000;
>> intvalinit('Display_')
>> for i=1:n;b(i,1)=-1;end;
>> diag_uncertainty = midrad(3,1e-4)
intval diag_uncertainty = 3.0000__________
>> for i=1:n;B(i,1) = midrad(-1,0);
B(i,2)=diag_uncertainty;
B(i,3) = midrad(-1,0);end
>> d = [-1;0;1];
>> A = spdiags(B,d,n,n);
>> A(1:3,1:3)
intval ans =
   (1,1)                    3.0000__________
   (2,1)                   -1.00000000000000
   (1,2)                   -1.00000000000000
   (2,2)                    3.0000__________
   (3,2)                   -1.00000000000000
   (2,3)                   -1.00000000000000
   (3,3)                    3.0000__________
>> X=verifylss(A,b);
>> X(1:4)
intval ans =
  -0.62____________
  -0.85____________
  -0.94____________
  -0.98____________
>> X(500:503)
intval ans =
  -1.00____________
  -1.00____________
  -1.00____________
  -1.00____________
>> X(997:1000)
intval ans =
  -0.98____________
  -0.94____________
  -0.85____________
  -0.62____________
```

The response and solution process in this interactive dialogue were almost immediate (small fractions of a second). □

Material on mathematically rigorous bounds to solutions of sparse systems of equations can be found in [221] and [226].

7.8 Final Notes

Mayer, Alefeld, Rohn, and others have made significant advances in characterizing the solution sets to interval linear systems. These results include characterizing necessary and sufficient conditions for when interval Gaussian elimination and other algorithms for bounding solution sets can be carried out and when they give sharp enclosures for the solution sets. They also include results for interval matrices with assumed dependency in their entries, such as for symmetric, persymmetric, Toeplitz, and Hankel matrices. Günter Mayer provided an excellent review of this in [140].

Jiri Rohn, aside from his role as a leader in these developments and in the analysis of computational complexity in interval numerical linear algebra, has made available a sizeable suite of INTLAB programs for interval linear systems of equations. He has written a concise summary of results on interval linear problems. See Appendix D.

In the next chapter, the ideas and techniques discussed in Chapters 6 and 7 are used to develop methods for mathematically rigorous bounds on the solutions to nonlinear systems of equations.

Chapter 8

Interval Newton Methods

In Chapter 6, we discussed iterative interval methods for solving equations in *fixed-point form*. Given an equation

$$x = f(x), \tag{8.1}$$

we take an interval extension F of f (which is automatically inclusion isotonic if f is rational) and set up an iterative procedure of the form

$$X_{k+1} = F(X_k) \cap X_k \qquad (k = 0, 1, 2, \ldots). \tag{8.2}$$

If we start with an X_0 such that $F(X_0) \subseteq X_0$, then (8.2) produces a nested sequence of intervals $\{X_k\}$ convergent to an interval X^* such that $X^* = F(X^*)$ and $X^* \subseteq X_k$ for all $k = 0, 1, 2, \ldots$. On a computer, the procedure can be halted when $X_{k+1} = X_k$; using IA at a specific number of digits, this yields the narrowest interval possible (with that many digits) containing X^*.

The Krawczyk method, considered in Chapter 6 in the linear case, falls under the same general scheme as (8.2) and can be used to solve nonlinear systems of equations. More generally, interval Newton methods share properties with the Krawczyk method, can be implemented with iteration (8.2), and can be used to prove existence and uniqueness of a solution to a nonlinear system of equations in a given box, even though the interval Newton operator is not inclusion isotonic as is F in (8.2). We describe these methods below.

8.1 Newton's Method in One Dimension

Our approach will be to initially develop Newton's method in its simplest form. Let f be a real-valued function of a real variable x, and suppose that f is continuously differentiable. By the mean value theorem, we can write

$$f(x) = f(y) + f'(s)(x - y) \tag{8.3}$$

for some s between x and y. Now let $[a, b]$ be an interval in which we seek a solution of the equation

$$f(x) = 0. \tag{8.4}$$

A solution x, if it exists, would satisfy

$$f(y) + f'(s)(x - y) = 0 \tag{8.5}$$

for any $y \in [a, b]$, in particular for $y = m([a, b]) = (a + b)/2$. Hence,

$$x = y - f(y)/f'(s). \tag{8.6}$$

Let $F'(X)$ be an inclusion monotonic interval extension of $f'(x)$ and consider the algorithm

$$X^{(k+1)} = X^{(k)} \cap N(X^{(k)}) \qquad (k = 0, 1, 2, \ldots), \tag{8.7}$$

where

$$N(X) = m(X) - f(m(X))/F'(X). \tag{8.8}$$

It follows from (8.6) that x is contained in $N(X)$ if $y = m(X)$, and if x is contained in X, then s in (8.3) is also contained in X. In this case, x is also contained in $X^{(k)}$ for all k if it is contained in $X^{(0)}$. We have the following theorem.

Theorem 8.1. If an interval $X^{(0)}$ contains a zero x of $f(x)$, then so does $X^{(k)}$ for all $k = 0, 1, 2, \ldots$, defined by (8.7). Furthermore, the intervals $X^{(k)}$ form a nested sequence converging to x if $0 \notin F'(X^{(0)})$.

Proof. If $0 \notin F'(X^{(0)})$, then $0 \notin F'(X^{(k)})$ for all k and $m(X^{(k)})$ is not contained in $N(X^{(k)})$, unless $f(m(X^{(k)}) = 0$. Therefore $w(X^{(k+1)}) < \frac{1}{2}w(X^{(k)})$. Convergence of the sequence follows. □

Exercise 8.1. What happens if $f(m(X^{(k)}) = 0$? □

Example 8.1. Suppose we wish to solve (8.4) with

$$f(x) = x^2 - 2. \tag{8.9}$$

An interval extension of $f'(x) = 2x$ is $F'(X) = 2X$. Hence,

$$N(X) = m(X) - \frac{[m(X)]^2 - 2}{2X},$$

and (8.7) looks like

$$X^{(k+1)} = X^{(k)} \cap \left\{ m(X^{(k)}) - \frac{[m(X^{(k)})]^2 - 2}{2X^{(k)}} \right\}.$$

Taking $X^{(0)} = [1, 2]$, we obtain

$$X^{(1)} = [\tfrac{22}{16}, \tfrac{23}{16}],$$
$$X^{(2)} = [1.41406\ldots, 1.41441\ldots],$$
$$X^{(3)} = [1.414213559\ldots, 1.414213566\ldots].$$

Of course, (8.4) has solution $x = \sqrt{2}$. Rounding $X^{(3)}$ out at the eighth place, we see that $\sqrt{2}$ lies in $[1.41421355, 1.41421357]$. □

The rapid convergence exhibited here is a consequence of the fact that the interval Newton method is *asymptotically error squaring*. A formal statement of this property is as follows.

Lemma 8.1. Given a real rational function f of a single real variable x with rational extensions F, F' of f, f', respectively, such that f has a simple zero y in an interval $[x_1, x_2]$ for which $F([x_1, x_2])$ is defined and $F'([x_1, x_2])$ is defined and does not contain zero, there is an interval $X_0 \subseteq [x_1, x_2]$ containing y and a positive real number C such that

$$w(X^{(k+1)}) \le C(w(X^{(k)}))^2.$$

For a proof, see Moore [148]. A simple illustration of the quadratic convergence of the interval Newton method compared to the quadratic convergence of the traditional point Newton method appears in [106].

Geometric Interpretation

The univariate interval Newton method has a geometric interpretation similar to that of the classical point Newton method. However, instead of the intersection of a single tangent line with the x-axis, the new interval $X^{(k+1)}$ is defined by the intersection of two tangent lines, with slopes corresponding to the lower and upper bounds of $F'(X^{(k)})$. This is illustrated with $F'(X^{(k)})$ equal to the range of f' over $X^{(k)}$ in Figure 8.1. (In this figure, the blue slopes represent the lower bounds on $F'(X^{(k)})$, and the red slopes represent the upper bounds on $F'(X^{(k)})$.)

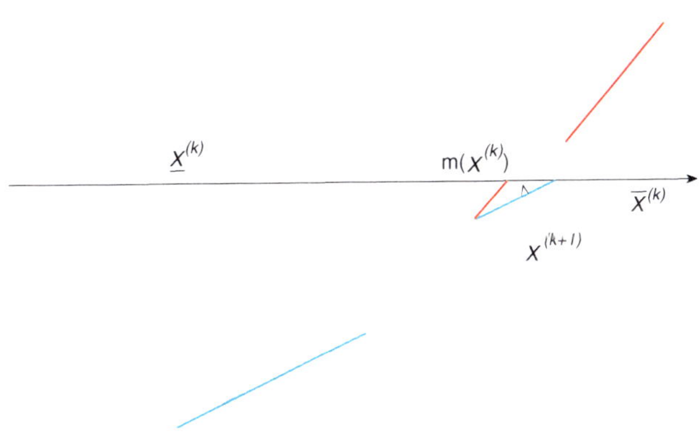

Figure 8.1. *Geometrical interpretation of the univariate interval Newton method.*

Exercise 8.2. Show that $X^{(k)} = X^{(3)}$ for all $k > 3$ in Example 8.1. That is, show that we have *finite convergence* to eight places in three iterations. □

If the intersection $X^{(k)} \cap N(X^{(k)})$ in (8.7) is empty, there is no zero of f in $X^{(k)}$.

Exercise 8.3. Illustrate this by taking $X^{(0)} = [4, 6]$ in Example 8.1. □

Example 8.2. The iteration equation (8.7) can be implemented easily with INTLAB in the following m-file `i_newton_step.m`:

```
function [NX_intersect_X, is_empty] = i_newton_step(f,f_prime,X)
% [NX_intersect_X, is_empty] = i_newton_step(f,f_prime,X)
%  returns the result of a single step of the interval Newton
%  method for a single variable, as defined in (8.8) of
%  the text, using X as initial interval.  The string f should
%  be the name of an "m" file for evaluating the function,
%  while the string "f_prime" should be the name of an  "m"
%  file for evaluating the derivative of f.  The flag "is_empty"
%  is set to "0" if the intersection is non-empty, and is set
%  to "1" if the intersection is empty.

midX = infsup(mid(X),mid(X));
NX = midX - feval(f,midX) / feval(f_prime,X)
NX_intersect_X = intersect(NX,X);
is_empty = isempty_(NX_intersect_X);
```

The function in Example 8.1 can be implemented in the following two m-files:

```
function [y] = example8p1(x);
% y = example8p1(x) returns x^2 -2 in y, regardless of the
% variable type of x;  y will be of the same variable type
% as x.
y = x.^2 - 2;
```

and

```
function [y] = example8p1_prime(x);
% y = example8p1(x) returns the derivative of x^2 -2 in y,
% regardless of the variable type of x;  y will be of the
% same variable type as x.
y = 2*x;
```

With these three m-files, we may produce the following MATLAB console dialogue:

```
>> format long
>> X = infsup(1,2)
intval X = [ 1.00000000000000 , 2.00000000000000 ]
>> [X,is_empty] = i_newton_step('example8p1','example8p1_prime',X)
intval X = [ 1.37499999999999 , 1.43750000000001 ]
is_empty = 0
>> [X,is_empty] = i_newton_step('example8p1','example8p1_prime',X)
intval X = [ 1.41406249999999 , 1.41441761363637 ]
is_empty = 0
>> [X,is_empty] = i_newton_step('example8p1','example8p1_prime',X)
intval X = [ 1.41421355929452 , 1.41421356594718 ]
is_empty = 0
```

```
>> [X,is_empty] = i_newton_step('example8p1','example8p1_prime',X)
intval X = [ 1.41421356237309 , 1.41421356237310 ]
is_empty = 0
>> [X,is_empty] = i_newton_step('example8p1','example8p1_prime',X)
intval X = [ 1.41421356237309 , 1.41421356237310 ]
is_empty = 0
```
□

Exercise 8.4. It is important for mathematical rigor to define `midX` as `infsup(mid(X), mid(X))` rather than simply as `mid(X)`. Explain. □

With some extra knowledge of the toolboxes distributed with INTLAB, there is no need to explicitly program the derivative, since the "gradient" package, supplying *automatic* (or *computational*) *differentiation* capabilities within INTLAB, may be used. (One of the first surveys of automatic differentiation is [200]. There have been many conference proceedings and several comprehensive books on the subject since then, and software packages in various programming languages have been written. For a short didactic explanation from among our own work, see [97, pp. 36–41]. Additional explanation of automatic differentiation appears in section 9.3 of this book, where we use the technique in computing integrals.) This results in the following m-file:

```
function [NX_intersect_X, is_empty] = i_newton_step_no_fp(f,X)
% [NX_intersect_X, is_empty] = i_newton_step(f,f_prime,X)
%  returns the result of a single step of the interval Newton
%  method for a single variable, as defined in (8.6) of
%  the text, using X as initial interval.  The string f should
%  be the name of an "m" file for evaluating the function,
%  while the string "f_prime" should be the name of an  "m"
%  file for evaluating the derivative of f.  The flag "is_empty"
%  is set to "0" if the intersection is non-empty, and is set
%  to "1" if the intersection is empty.

midX = infsup(mid(X),mid(X));
Xg = gradientinit(X);
fXg = feval(f,Xg);
NX = midX - feval(f,midX) / fXg.dx;
NX_intersect_X = intersect(NX,X);
is_empty = isempty_(NX_intersect_X);
```

Exercise 8.5. Repeat the computation above Exercise 8.4 with `i_newton_step_no_fp` in place of `i_newton_step`. Are the same results displayed? (Note that, in general, the displayed results need not be the same, even if `format long` is used for both displays. Why do you think this is so?) □

Extended Interval Arithmetic

Let us continue to consider the function (8.9). After all, it has another zero, $-\sqrt{2}$. It would be nice if we could choose a sufficiently wide starting interval to encompass both zeros and find them both. However, for (8.9), a starting interval such as $[-2, 2]$ would violate the condition $0 \notin F'(X^{(0)})$ in the proof of Theorem 8.1. This situation can be handled using *extended interval arithmetic.*

The definition of interval division can be extended as follows. Recall that

$$[a, b]/[c, d] = [a, b](1/[c, d]),$$

where

$$1/[c, d] = \{1/y : y \in [c, d]\} \qquad (c < d \text{ any real numbers}).$$

If $0 \notin [c, d]$, we are using ordinary interval arithmetic. If $0 \in [c, d]$, extended interval arithmetic specifies three cases:[31]

1. If $c = 0 < d$, then $1/[c, d] = [1/d, +\infty)$.
2. If $c < 0 < d$, then $1/[c, d] = (-\infty, 1/c] \cup [1/d, +\infty)$.
3. If $c < d = 0$, then $1/[c, d] = (-\infty, 1/c]$.

We will not allow the case $c = d = 0$ here.

This definition makes the most sense if we keep in mind that the result of the operation should be the set of all possible values that can be obtained as we select numbers from the interval operand.

Exercise 8.6. Use extended interval arithmetic to compute $1/[-24, 3]$. □

With this, we can allow the range of values of the derivative $f'(x)$ to contain zero and the quotient $f(x)/f'(X)$ occurring in the computation of $N(X) = m(X) - f(x)/f'(X)$ to split into two unbounded intervals. Then, upon intersecting $N(X)$ with the finite interval X in the iteration formula

$$X^{(k+1)} = N(X^{(k)}) \cap X^{(k)},$$

we obtain two disjoint finite intervals. For (8.9), we could take $X^{(0)} = [-2, 2]$. We would obtain

$$m(X^{(0)}) = 0, \qquad f(m(X^{(0)})) = -2,$$

and

$$f'(X^{(0)}) = 2[-2, 2] = [-4, 4].$$

Hence,

$$\begin{aligned} N(X^{(0)}) &= m(X^{(0)}) - f(m(X^{(0)}))/f'(X^{(0)}) \\ &= 0 - \{-2/[-4, 4]\} \\ &= 1/[-2, 2] \\ &= \left(-\infty, -\tfrac{1}{2}\right] \cup \left[\tfrac{1}{2}, \infty\right). \end{aligned}$$

[31]These definitions can be derived according to Cset theory, which we briefly describe on p. 113. There is some variation, depending on whether $-\infty$ and ∞ are included in the underlying set. These definitions assume the underlying set is the real numbers, rather than the set of extended real numbers, including $-\infty$ and ∞. This has practical consequences. See note 4 on p. 115.

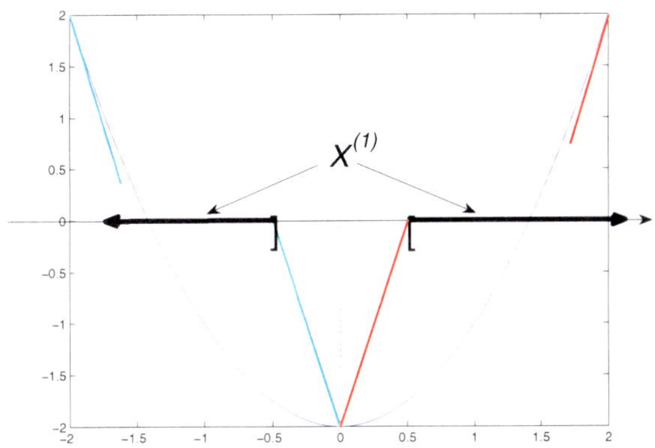

Figure 8.2. *Extended interval Newton step over $X^{(0)} = [-2, 2]$, Example 8.9.*

Intersecting this with $X^{(0)} = [-2, 2]$, we get the union of two disjoint finite intervals,

$$X^{(1)} = \left[-2, -\tfrac{1}{2}\right] \cup \left[\tfrac{1}{2}, 2\right].$$

This is illustrated in Figure 8.2.

We can momentarily put aside one of these two intervals, say, $\left[-2, -\frac{1}{2}\right]$, and set $X^{(0)} = \left[\frac{1}{2}, 2\right]$. After convergence for that case (which we have already seen), we can go back and set $X^{(0)} = \left[-2, -\frac{1}{2}\right]$ to find the other root in $[-2, 2]$. In this way, the interval Newton method can find all the zeros of a function in a given starting interval.

A more difficult case was carried out [154] for finding all the zeros of the function defined by

$$f(x) = x^2 + \sin(1/x^3) \text{ for } x \text{ in } [0.1, 1].$$

After repeated splittings and eventual convergence of all the generated subintervals, all 318 zeros in $[0.1, 1]$ were found, each within computed intervals of width less than 10^{-10}. They are packed very close together, especially near $x = 0.1$. The closest zero of f to that left endpoint of the starting interval $[0.1, 1]$, is contained in $[0.10003280626, 0.10003280628]$.

Exercise 8.7. Use the interval Newton method with extended arithmetic to find all zeros of the polynomial $x^3 - 5x - 1$. (You may corroborate your results with alternate techniques.) Start with an interval in which you know all roots lie. □

Exercise 8.8. It can be shown [234] that if a wooden sphere has radius 1 ft and specific gravity $\frac{2}{3}$, then the depth h (in ft) to which it will sink in water is given by $h^3 - 3h^2 + \frac{8}{3} = 0$. Find h. □

The interval arithmetic in INTLAB does not implement sharp extended arithmetic; division A/B by any interval B that contains zero results in the interval $[-\infty, \infty]$. This is a "correct" enclosure of the exact range of the operation but is in general not "sharp," since it contains many values that are not obtainable as a/b for $a \in A$ and $b \in B$. However, we may use the following function:

```
function [Y1,Y2,two] = xreciprocal(X)
%[Y1,Y2,two] = xreciprocal(X) returns the extended reciprocal
% of X defined by the three cases above Exercise 8.5 in the
% text.  The return value two is set to 0 if only one interval
% is returned and is set to 1 if two intervals are returned.
% If X does not contain zero, the result of ordinary
% interval division is returned in Y1, and two is set to 0.
% In the case inf(X) = sup(X) = 0, avoided in the text,
% two is set to 1, and two empty intervals  are returned.
% (INTLAB represents an empty interval as infsup(NaN,NaN) )
% In cases where there is only one interval, Y2 is set
% to INTLAB's representation of the empty interval.

if (inf(X) > 0) | (sup(X) < 0) % do ordinary interval division
    two=0;
    Y1 = 1/X;
    Y2 = infsup(NaN,NaN);
elseif (inf(X)==0) & (sup(X) > 0) % Case 1 of the text --
    two=0;
    lower_bound = infsup(1,1) / infsup(sup(X),sup(X));
    Y1 = infsup(inf(lower_bound),Inf);
    Y2 = infsup(NaN,NaN);
elseif (inf(X)<0) & (sup(X) > 0) % Case 2 of the text --
    two=1;
    upper_bound = infsup(1,1) / infsup(inf(X),inf(X));
    Y1=infsup(-Inf,sup(upper_bound));
    lower_bound = infsup(1,1) / infsup(sup(X),sup(X));
    Y2 = infsup(inf(lower_bound),Inf);
elseif (inf(X) < 0) & (sup(X) == 0) % Case 3 of the text --
    two = 0;
    upper_bound = infsup(1,1) / infsup(inf(X),inf(X));
    Y1=infsup(-Inf,sup(upper_bound));
    Y2 = infsup(NaN,NaN);
else % This is the case where X=0, not covered in the text --
    two =1;
    Y1 = infsup(NaN,NaN);
    Y2 = infsup(NaN,NaN);
end
```

The function `xreciprocal` can be used to automate the computations explained above for finding all roots of an equation. However, implementation requires storage of a list of intervals that have been set aside for later. This process is similar to the process for creating and using the list produced in the Skelboe–Moore algorithm through bisection.

Exercise 8.9. Redo Exercises 8.6, 8.7, and 8.8 using INTLAB and `xreciprocal`. □

Exercise 8.10. The function `xreciprocal` has various complications necessary to ensure that the results it returns contain the exact range of the reciprocation operation. Explain these complications. □

In case we wish to find intervals containing the roots of a rational function f which has coefficients known only to lie in certain intervals, we can take an interval extension F which evaluates the range of values of f over the intervals of coefficients and use the more general form

$$N(X) = m(X) - \frac{F(m(X))}{F'(X)} \tag{8.10}$$

for the Newton iteration function $N(X)$.

Example 8.3. To find the roots of $f(x) = x^2 - c$ that lie in the interval $[1, 2]$, for $c \in [2, 3]$, we can put

$$F(X) = X^2 - [2, 3], \quad F'(X) = 2X, \quad X_0 = [1, 2],$$

and find $N(X_0) \subseteq [1.39, 1.76]$. After one application of $X^{(k+1)} = N(X^{(k)}) \cap X^{(k)}$, we have

$$\{x : x^2 - c = 0, \; c \in [2, 3]\} \cap [1, 2] = [\sqrt{2}, \sqrt{3}] \subseteq [1.39, 1.76]. \qquad \square$$

A fundamental distinction between the interval Newton method and the ordinary Newton method is that the former uses *computation with sets* instead of computation with points. Again, this permits us to find *all* the zeros of a function in a given starting interval. Whereas the ordinary Newton method is prone to erratic behavior, the interval version practically always converges. The difference in performance of the two methods can be dramatic, and the interested reader is referred to Moore [152] for specific examples.

Exercise 8.11. The current I (mA) flowing in a diode circuit satisfies the equation $1 - I - 0.0052 \ln(10^9 I) = 0$. Solve by the Newton method. $\square$

Exercise 8.12. The odd-numbered transverse electric modes of a glass slab waveguide have permitted incidence angles θ_i given by [83],

$$\tan\left(\frac{3}{2}\pi \cos\theta_i\right) = \frac{\sqrt{\sin^2\theta_i - (9/4)^{-1}}}{\cos\theta_i}.$$

Find all solutions θ_i that lie in the interval $\left[\frac{\pi}{8}, \frac{\pi}{2}\right]$. (Be sure to choose starting intervals in such a way that singularities are avoided.) $\square$

Notes

1. Extended interval arithmetic involving reciprocation or division by intervals that contain zero has been the subject of much discussion and some controversy among experts in interval computation. In particular, various systems have been devised for doing computations on objects in which the system is closed under division by intervals that contain zero. The first such work, still widely referenced and the basis of what followed, is "Kahan arithmetic" [89]; the extended reciprocation defined here can be considered to contain elements of Kahan arithmetic. The extended reciprocation defined here serves well in the context of our univariate interval Newton algorithm. However, there are contexts when it is useful to represent both semi-infinite intervals resulting from the division as a single object, continuing to do nonstop, exception-free arithmetic on these objects. In such arithmetics, there are issues of efficiency

of the operations when using standards such as the IEEE 745 standard for binary floating point arithmetic, as well as issues of sharpness in enclosing exact ranges. There are also issues of consistency and mathematical elegance. For instance, some analyses suggest that it is preferable to have a system where, when $c = 0 < d$, the set $\{-\infty\} \cup [1/d, \infty]$ is returned instead of $[1/d, \infty]$. In such arithmetics, cases 1 and 3 of the reciprocation described above are limiting cases of case 2. A recent work discussing this is [194]. In this work, a consistent mathematical system, *Cset arithmetic*, is developed. The *Cset*, or "containment set" of an arithmetic expression of n variables at a point in $\mathbb{R}^n$ is the set of all limits of the expression as the variables in the expression tend to the point, and the Cset evaluation of a set or interval vector is the union of the Csets of the points in the set. The interval evaluation of an expression over an interval vector is then defined to be an enclosure of the Cset of the expression over that interval vector.

Cset arithmetic is compelling from the points of view of simplicity and connections with classical mathematical analysis. It also provides an unambiguous theoretical guide about how the arithmetic operations and function definitions (exponentials, trigonometric functions, etc.) should be extended.

2. The representations in function `xreciprocal` are presently subject to discussion as part of an effort to standardize interval arithmetic, and the ultimate standardized representation may differ from that seen in `xreciprocal`. For example, some view $1/0$ as the set $[-\infty, -\infty] \cup [\infty, \infty]$, rather than the empty set; furthermore, ultimately, the empty set might be represented by something other than `[NaN,NaN]`.

3. A related controversial issue is whether the underlying set is the set of real numbers or the set of extended real numbers, including $-\infty$ and ∞. As an example of the difference, consider applying the interval Newton method to the function $f(x) = \sin(\pi x)/\pi$ with starting interval $X = [-2.5, 2.5]$, so $m(X) = 0$, while the exact range of f' over X is $[-1, 1]$. Thus, the iteration (8.10) becomes

$$N(X) = 0 - \frac{0}{[-1, 1]}.$$

If we take the underlying model to be the set of real numbers but interpret $0/0$ according to Cset theory to be the set of all limits of x/y as $x \to 0$ and $y \to 0$, then we obtain $0/0 = (-\infty, \infty)$, so $N(X) = (-\infty, \infty)$, and $N(X)$ contains all solutions of $f(x) = 0$ in X. (The solutions are $x = -2, -1, 0, 1$, and 2.) On the other hand, suppose we write $N(X)$ as

$$\tilde{N}(X) = m(X) - F(m(X)) \cdot \frac{1}{F'(X)}.$$

Assuming the underlying model is the set of real numbers, and not the extended reals, we have $1/F'(X) = 0 \cdot (-\infty, \infty) = 0$, and $\tilde{N}(X) = [0, 0]$. Thus, in this case, $\tilde{N}(X)$ does not contain all solutions to $f(x) = 0$ that are in X, that is, Theorem 8.1 is not true with this extension of interval arithmetic. However, these operational definitions of $0/[-1, 1]$ and $0 \cdot (-\infty, \infty)$ are implemented in many currently available and proposed systems for doing interval arithmetic.

In contrast, if we assume the two-point compactification of $\mathbb{R}$, that is, if we assume that $-\infty$ and ∞ are numbers in the system, then the Cset of $0/[-1, 1]$ is $[-\infty, \infty]$, while the Cset of the product $0 \cdot [-\infty, \infty]$ must contain the Csets of $0 \cdot -\infty$ and $0 \cdot \infty$. However, the set of all limits of $x \cdot y$ as $x \to 0$ and $y \to \pm\infty$ is $[-\infty, \infty]$, so $N(X) = \tilde{N}(X) = [-\infty, \infty]$, and it is not critical how $N(X)$ is implemented.

As of the writing of this book, INTLAB returns `[NaN,NaN]` for each of $0/[-1, 1]$ and $0 \cdot [-\inf, \inf]$ but returns $[-\inf, \inf]$ for $1/[-1, 1]$. In this context, `[NaN,NaN]` can be interpreted to mean "an undefined result."

4. A final issue related to interval Newton methods and to the Krawczyk method we present in the next section is how to define interval extensions of functions over arguments that contain both points inside the domain of the function and points outside the domain of the function. To study this, we note that a simplification of the Brouwer fixed-point theorem[32] states that any continuous mapping g of a closed and bounded interval X into itself has the property that there is an $x \in X$ with $g(x) = x$. Thus, if we take an interval evaluation of g over X and the resulting interval $g(X)$ has $g(X) \subset X$, we can conclude that there is an $x \in X$ such that $g(x) = x$. However, consider $g(x) = \sqrt{x} - 3$ and $X = [-4, 4]$. If we define $g(X)$ to be an interval that contains the range of g over that portion of X in the domain of g, then the narrowest such interval will be $g(X) \subseteq [-3, -1]$. Since $g(X) \subset X$, we would come to the false conclusion that there is a point $x \in X$ such that $g(x) = x$. We can explain this by noting that g is not continuous over $[-2, 2]$.

   In contrast, suppose we knew that $x \in [-4, 4]$ and that $x = (x + 3)^2$, and we wanted to find a narrower interval than $[-4, 4]$ in which x must lie, if such an x exists in $[-4, 4]$. (Such computations are important in various systems found in practice, involving *constraint propagation*.) We would then have $x = \sqrt{x} - 3$, or $x = -\sqrt{x} - 3$, and the above computation correctly gives $x \in [-3, -1] \cap [-4, 4] = [-3, -1]$ or $x \in [-5, -3] \cap [-4, 4] = [-4, -3]$, whence $x \in [-4, -1]$, a correct result. (We may then plug $[-4, -1]$ into the original equation to obtain $x \in ([-4, 1] + 3)^2 = [-1, 4]^2 = [0, 16]$, and, since $[-3, -1] \cap [0, 16] = 0$, we conclude that there is no $x \in [-4, 4]$ satisfying $x = (x + 3)^2$.)

   Thus, evaluation of an interval extension over an interval that is partially out of the domain will not lead to a practical result in the case of proving existence of solutions but does lead to a practical result in the case of constraint propagation. One proposal foı handling this problem in implementations of interval arithmetic is with a "discontinuous" flag[33] that is raised when functions are evaluated over intervals that lie partially outside their domains.

   After issuing a warning, INTLAB returns the complex interval $[-2, 2] + [-2, 2]i$ for $\sqrt{[-2, 2]}$, representing a third possible context of use.

[32] In fact, the Krawczyk method, described below, can be analyzed in terms of the Brouwer fixed-point theorem.

[33] Similar to the "inexact" flag and other flags that are raised in systems conforming to the IEEE 754 standard for binary floating point arithmetic.

8.2 The Krawczyk Method

In this and the following sections, we consider finite systems of nonlinear equations

$$\begin{aligned} f_1(x_1, \ldots, x_n) &= 0, \\ &\vdots \\ f_n(x_1, \ldots, x_n) &= 0, \end{aligned} \tag{8.11}$$

which we may write in vector notation as

$$f(x) = 0. \tag{8.12}$$

In this section, we consider the Krawczyk method (first introduced for linear systems as (7.7)), while in section 8.4 we consider a somewhat more straightforward generalization of the univariate interval Newton method. Each approach has its own advantages, and both are used in practice.

We can consider two cases: (1) the functions f_i are exactly representable real-valued functions, or (2) the functions f_i have coefficients known only to lie in certain intervals. We will discuss case (1) first and extend the results to case (2).

Suppose that f in (8.12) is continuously differentiable in an open domain D. Suppose that we can compute inclusion isotonic interval extensions F and F' for f and f', defined on interval vectors $X \subseteq D$. We have the following computational test for the existence of a solution [119, 150].

Theorem 8.2. Let Y be a nonsingular real matrix approximating the inverse of the real Jacobian matrix $F'(m(X))$ with elements $F'(m(X))_{ij} = \partial f_i(x)/\partial x_j$ at $x = m(X)$. Let y be a real vector contained in the interval vector $X \subseteq D$. Define $K(X)$ by

$$K(X) = y - Yf(y) + \{I - YF'(X)\}(X - y). \tag{8.13}$$

If $K(X) \subseteq X$, then (8.12) has a solution in X. It is also in $K(X)$.

An early work containing the proof of this theorem is [150]. The proof is based on a generalization of the Brouwer fixed-point theorem, or a specific instance of the Schauder fixed-point theorem. Namely, if X is a closed convex subset of $\mathbb{R}^n$, and g maps X into itself, then g has a fixed point x^*, $x^* = g(x^*)$, $x^* \in X$.

Proof (of Theorem 8.2). Define $g(y) = y - Yf(y)$. Then, since Y is nonsingular, $g(y) = y$ if and only if $f(y) = 0$. Thus, if there exists an $x \in X$ such that $g(x) = x$, that is, if there is a fixed point of g in x, then there is a solution to $f(x) = 0$ in x. However, if by $g'(y)$ we mean the Jacobian matrix of g at y, then

$$g'(y) = I - Yf'(y).$$

Therefore, the mean value extension (see (6.25)) of g over X about the point $y \in X$ is simply[34]

$$y - Yf(y) + \{I - YF'(X)\}(X - y) = K(X).$$

[34] Here, we have (6.25) for each component of f.

Thus, $K(X)$ must contain the range of g over X, that is, $g(X) \subseteq K(X)$. Thus, if $K(X) \subseteq X$, then $g(X) \subseteq X$, the hypotheses of the Schauder fixed-point theorem hold, so g has a fixed point in X, so $f(x) = 0$ has a solution in X. □

If the interval vector $X = (X_1, \ldots, X_n)$ is an n-cube so that $w(X_i) = w(X)$ for $i = 1, \ldots, n$, and if we choose $y = m(X)$, then $K(X)$ lies in the interior of X if

$$\|K(X) - m(X)\| < w(X)/2. \tag{8.14}$$

Thus, for an n-cube X, (8.14) is sufficient for the existence of a solution to (8.12) in X. The same condition (8.14), which can be verified computationally by evaluating $K(X)$, is also sufficient to guarantee convergence of the interval Krawczyk method.

Theorem 8.3. Let X be an n-cube, $y = m(X)$, and Y a nonsingular real matrix. Suppose (8.14) is satisfied. Put $X^{(0)} = X$, $Y^{(0)} = Y$ and consider an arbitrary real vector $x^{(0)}$ in $X^{(0)}$. Then the system (8.11) has a unique solution in X, and the following algorithm converges to the solution [150, 151]:

$$X^{(k+1)} = X^{(k)} \cap K(X^{(k)}) \qquad (k = 1, 2, \ldots), \tag{8.15}$$

where

$$K(X^{(k)}) = y^{(k)} - Y^{(k)} f(y^{(k)}) + \{I - Y^{(k)} F'(X^{(k)})\} Z^{(k)}$$

and

$$y^{(k)} = m(X^{(k)}), \qquad Z^{(k)} = X^{(k)} - m(y^{(k)}),$$

and where $Y^{(k)}$ is chosen as

$$Y^{(k)} = \begin{cases} Y, \text{ an approximation to } [m(F'(X^{(k)}))]^{-1} \\ \text{if } \left\|I - YF'(X^{(k)})\right\| \le \left\|I - Y^{(k-1)} F'(X^{(k-1)})\right\|, \\ Y^{(k-1)} \text{ otherwise.} \end{cases}$$

Example 8.4. Consider the system of equations

$$\begin{aligned} f_1(x_1, x_2) &= x_1^2 + x_2^2 - 1 = 0, \\ f_2(x_1, x_2) &= x_1 - x_2^2 = 0. \end{aligned} \tag{8.16}$$

For f' we have the Jacobian matrix

$$f'(x) = \begin{pmatrix} 2x_1 & 2x_2 \\ 1 & -2x_2 \end{pmatrix}. \tag{8.17}$$

For the interval extensions F and F', we can take the natural interval extensions of the corresponding real rational functions, simply evaluating (8.16) and (8.17) in interval arithmetic:

$$\begin{aligned} F_1(X) &= X_1^2 + X_2^2 - 1, \\ F_2(X) &= X_1 - X_2^2, \end{aligned} \qquad \text{and} \qquad F'(X) = \begin{pmatrix} 2X_1 & 2X_2 \\ 1 & -2X_2 \end{pmatrix}.$$

Suppose we decide to try $X = ([0.5, 0.8], [0.6, 0.9])^T$. Then, we have

$$m(F'(X)) = \begin{pmatrix} m([1, 1.6]) & m([1.2, 1.8]) \\ m([1, 1]) & m([-1.8, -1.2]) \end{pmatrix} = \begin{pmatrix} 1.3 & 1.5 \\ 1 & -1.5 \end{pmatrix}.$$

As an approximate inverse of this matrix, we will take

$$Y = \begin{pmatrix} 0.43 & 0.43 \\ 0.29 & -0.37 \end{pmatrix}. \tag{8.18}$$

Putting $y = (0.65, 0.75)^T = m(X)$, we find from (8.13) for the 2-cube X,

$$K(X) \subseteq ([0.559, 0.68], [0.74, 0.84])^T.$$

Since $\|K(X) - m(X)\| = 0.091 < w(X)/2 = .15$, the hypotheses for Theorem 8.3 are satisfied. The iterative method (8.15) converges to a solution of (8.16) using Y given by (8.18) from $X^{(0)} = ([0.5, 0.8], [0.6, 0.9])^T$. It produces a nested sequence of interval vectors containing the solution, and using IA converges in a finite number of steps to an interval vector containing a solution of (8.16). □

It can be shown [159] that the widths of the containing interval vectors converge *quadratically* to zero if $F'(X)$ is a Lipschitz extension of $f'(x)$.

INTLAB Implementation

The Krawczyk method can be implemented in INTLAB with the following m-file:

```
function [KX] = Krawczyk_step(X,y,f)
% [KX] = Krawczyk_step(X,y,f,fprime) returns the image of X
% from a step of the Krawczyk method, with base point y, where
% the function is programmed in the function whose name is in
% the string f.  There is no need to program the Jacobian matrix,
% since the "gradient" automatic differentiation toolbox,
% distributed with INTLAB, is used.  It is the user's
% responsibility to ensure that X is an interval column vector,
% that y is a non-interval column vector contained in X, and
% that f is the desired function in an "m" file that returns
% a column vector.

% First compute f(y) using interval arithmetic to bound
% roundoff error --
n=length(X);
iy = midrad(y,0);
fy = feval(f,iy);

% Now compute F'(X) and the preconditioning matrix Y --
Xg = gradientinit(X);
FXg = feval(f,Xg);

Y = inv(mid(FXg.dx));
```

```
% Finally, compute the actual Krawczyk step --
KX = y - Y*fy + (eye(n,n) - Y*FXg.dx) * (X - y);
```

Since INTLAB's `gradient` package is used in `Krawczyk_step.m`, there is no need to explicitly program the Jacobian matrix. To use `Krawczyk_step` to do the computations above for Example 8.4, we supply the following m-file:

```
function [f] = example8p4(x)
% [fx] = example8p4(x) returns the components f_1 and f_2
% for Example 8.4 of the text in f.  It is important that
% the return value f be a column vector.  (If f is defined
% by componentwise assignment f(1) = ..., f(2) = ..., then
% Matlab represents f as a row vector by default, and one
% must take f=f'.)

f = [x(1)^2 + x(2)^2 - 1
     x(1) - x(2)^2];
```

The MATLAB command window dialogue can then be as follows:

```
>> X = [infsup(.5,.8)
        infsup(.6,.9)]
intval X =
[ 0.50000000000000 , 0.80000000000001 ]
[ 0.59999999999998 , 0.90000000000001 ]
>> y = mid(X)
y = 0.650000000000000
    0.750000000000000
>> Kx = Krawczyk_step(X,y,'example8p4')
intval Kx =
[ 0.55978260869565 , 0.67717391304348 ]
[ 0.74427536231884 , 0.83036231884059 ]
```

Exercise 8.13. Using INTLAB, repeat Example 8.4 with f_2 replaced by $f_2(x_1, x_2) = x_1 - x_2$. □

Exercise 8.14. An electric circuit is described by the equations

$$20 - 20I - V = 0, \qquad I - \frac{V}{20} - 10^{-9}e^{V/0.052} = 0.$$

Find I and V using INTLAB. □

Exercise 8.15. The state (x_1, x_2) of a certain nonlinear system is described by the differential equations [110]

$$\frac{dx_1}{dt} = 0.5[-h(x_1) + x_2], \qquad \frac{dx_2}{dt} = 0.2(-x_1 - 1.5x_2 + 1.2),$$

where $h(x_1) = 17.76\,x_1 - 103.79\,x_1^2 + 229.62\,x_1^3 - 226.31\,x_1^4 + 83.72\,x_1^5$. Use the Krawczyk method within INTLAB to find the system's critical points. □

Exercise 8.16. If the semicolon is removed from the end of a line in an m-file, then MATLAB prints the result of the computation on that line in the command window. (If there is more than one computation on a line, the separating semicolons can be replaced by commas to have MATLAB print those results in the command window.) Remove semicolons from `Krawczyk_step.m` and run it, so you can see the intermediate results displayed above for Example 8.4. Observe how `y`, `fy`, and `FXg.dx` enter into the formula as you iteratively call `Krawczyk_step` several times. □

Krawczyk originally proposed his method as an iterative method for obtaining narrower bounds on a solution, once initial bounds were known. Later, Moore derived the norm-based existence test in Theorem 8.3. Afterward, in [218], Rump showed that $K(X) \subset X$ implied there is a unique solution within X, without needing to compute matrix norms; also in [218], an "epsilon-inflation" process, commonly used today, was introduced. Epsilon-inflation is used to obtain both tight rigorous error bounds and a large region in which a solution is unique, given an approximate solution. Rump has embodied these ideas in the m-file `verifynlss`, supplied with INTLAB. `verifynlss` uses floating point iteration and Krawczyk's method to compute bounds and verify existence and/or uniqueness of a solution to a system of equations within those bounds; it begins with a floating point guess for the solution. For details of the mathematical background for `verifynlss`, see [219] and [221]. To use `verifynlss` in Example 8.4, we might have the following command window dialogue:

```
>> Solution = verifynlss('example8p4',[.65, .75],'g')
intval Solution =
[ 0.6180 , 0.6181 ]
[ 0.7861 , 0.7862 ]
>> format long
>> Solution
intval Solution =
[ 0.61803398874989 , 0.61803398874990 ]
[ 0.78615137775742 , 0.78615137775743 ]
```

Exercise 8.17. Try using `verifynlss` to find narrow bounds in which each solution in Example 8.4, Exercise 8.14, and Exercise 8.15 are proven to be unique. □

Exercise 8.18. Repeat Exercise 8.17 with the simpler routine `Krawczyk_step`. (This will probably involve calling `Krawczyk_step` repeatedly.) □

Exercise 8.19. Both `Krawczyk_step` and `verifynlss` can be used directly for the univariate examples in section 8.1. For instance, `'example8p1'` can be used as an argument for either of these, without modifying `example8p1.m`. Redo Exercise 8.8, Example 8.3, and Exercise 8.11 from section 8.1 using `Krawczyk_step` and using `verifynlss`. □

In case (2) mentioned at the beginning of this section, when the functions f_i in (8.11) have coefficients known only to lie in certain intervals, we can still use the algorithm (8.15). In this case, we require inclusion isotonic interval functions F and F' such that $f(x)$ is contained in $F(x)$ for every choice of real constants in the interval coefficients and such that $f'(x)$ is contained in $F'(X)$ for every choice of real constants from the interval coefficients and for every x in X. We replace $f(y^{(k)})$ in (8.15) by $F(y^{(k)})$, and the condition (8.14) now implies that the system (8.11) has a solution in X (perhaps a set of them) and that the sequence of interval vectors generated by (8.15), starting with $X^{(0)} = X$, $Y^{(0)} = Y$, is a nested sequence and converges in a finite number of steps (using IA) to an interval vector containing all the solutions to (8.11) in X for any choice of the real coefficients from the given interval coefficients. Thus we can compute bounds on the set of possible solutions in X.

8.3 Safe Starting Intervals

A *safe starting interval* for the interval Newton method is an $X^{(0)}$ from which the method (8.15) converges to a solution of a given nonlinear system of the form (8.11) or (8.12). From Theorem 8.3, $X^{(0)}$ is a safe starting interval if (8.14) holds; equivalently if $K(X^{(0)}) \subset X^{(0)}$, that is, if $K(X^{(0)})$ is contained in the interior of $X^{(0)}$,

$$K(X^{(0)}) \subset X^{(0)} \implies \text{a solution is in } K(X^{(0)}). \tag{8.19}$$

From Theorem 8.2, with $K(X)$ defined by (8.13), if $K(X) \subseteq X$, there is a solution in X, and it is also in $K(X)$, so there is a solution in the intersection $K(X) \cap X$. It follows that if $K(X) \cap X$ is empty, there is no solution in X:

$$K(X) \cap X = \emptyset \implies \text{no solution in } X. \tag{8.20}$$

In searching within a region B of interest, where we would like to find solutions, we can delete any part of B for which (8.20) holds. Furthermore, the extended definition of interval division discussed following Exercise 8.3, which splits an interval and produces a gap containing no solutions, can be extended to systems of nonlinear equations.

Example 8.5. We reconsider Exercise 8.12 to illustrate how we can find safe starting intervals. We are interested in finding all solutions that lie in the interval $[\frac{\pi}{8}, \frac{\pi}{2}]$ to the equation

$$\tan\left(\frac{3}{2}\pi \cos\theta_i\right) = \frac{\sqrt{\sin^2\theta_i - (9/4)^{-1}}}{\cos\theta_i}.$$

We can rewrite the equation in the form $f(x) = 0$, using x instead of θ_i, where

$$f(x) = \cos(x)\tan\left(\tfrac{3}{2}\pi\cos x\right) - \sqrt{(\sin x)^2 - \tfrac{4}{9}}.$$

The problem presents the interesting challenge that $f(x)$ is not defined on all of the interval of interest $\left[\frac{\pi}{8}, \frac{\pi}{2}\right]$.

We can find the expression for

$$K(X) = y - Yf(y) + \{I - YF'(X)\}(X - y)$$

for this example as follows. For the derivative of $f(x)$, we find that

$$f'(x) = -\tan\left(\tfrac{3}{2}\pi\cos x\right)(\sin x) - \tfrac{3}{2}\pi\frac{\sin x}{\cos^2\left(\frac{3}{2}\pi\cos x\right)}(\cos x)$$
$$- 3\frac{\sin x}{\sqrt{5 - 9\cos^2 x}}\cos x. \tag{8.21}$$

We cannot evaluate the natural interval extension of $f'(x)$ for the interval $X = \left[\frac{\pi}{8}, \frac{\pi}{2}\right]$ for several reasons.

First, the term $\sqrt{5 - 9\cos^2 x}$ defines only a real-valued function for values of x such that $5 - 9\cos^2 x \geq 0$. This will be the case only if $\cos^2 x \leq \frac{5}{9}$. This already restricts x to a subinterval of $\left[\frac{\pi}{8}, \frac{\pi}{2}\right] = [0.3926990816\ldots, 1.570796326\ldots]$. In fact, we are restricted to values of x in the subinterval (of $\left[\frac{\pi}{8}, \frac{\pi}{2}\right]$) given by

$$x \in \left[\cos^{-1}\left(\sqrt{\tfrac{5}{9}}\right), \tfrac{\pi}{2}\right] = [0.7297276562\ldots, 1.570796326\ldots].$$

Furthermore,

$$\tan\left(\tfrac{3}{2}\pi \cos x\right) = \sin\left(\tfrac{3}{2}\pi \cos x\right) / \cos\left(\tfrac{3}{2}\pi \cos x\right),$$

so both the first and second terms in (8.21) have $\cos(\frac{3}{2}\pi \cos x)$ in the denominator, and there is a division by 0 in those two terms for x such that $\frac{3}{2}\pi \cos x = \frac{\pi}{2}$. This happens for $x = \cos^{-1}(\frac{1}{3}) = 1.230959417\ldots$, which is in our interval of interest. Thus we can only evaluate an interval extension of (8.21) for proper subintervals of one or the other of the two intervals $[\cos^{-1}(\sqrt{\frac{5}{9}}), \cos^{-1}(\frac{1}{3})]$ and $[\cos^{-1}(\frac{1}{3}), \frac{\pi}{2}]$. To 10 places, these are the intervals

$$\left[\cos^{-1}\left(\sqrt{\tfrac{5}{9}}\right), \cos^{-1}\left(\tfrac{1}{3}\right)\right] = [0.7297276562\ldots, 1.230959417\ldots]$$

and

$$\left[\cos^{-1}\left(\tfrac{1}{3}\right), \tfrac{\pi}{2}\right] = [1.230959417\ldots, 1.570796326\ldots].$$

In fact, Aberth's RANGE found the zero $x = 0.76551 \pm 5(10^{-6})$ starting with $X = [0.73, 1.2]$ and the zero $x = 1.309585 \pm 5(10^{-7})$ starting with $X = [1.25, 1.57]$ and found that there are no further zeros in those starting intervals, using tests such as (8.19) and (8.20) above. □

For any given trial box $X^{(0)}$, there are various things that can happen, including the following:

1. $K(X^{(0)}) \subset \text{int}(X^{(0)}) \implies$ convergence to the unique solution in $X^{(0)}$, where $\text{int}(X^{(0)})$ denotes the topological interior of the box $X^{(0)}$.

2. $K(X^{(0)}) \cap X^{(0)} = \emptyset \implies$ no solution in $X^{(0)}$.

3. $K(X^{(0)}) \cap X^{(0)} \neq \emptyset \implies$ no conclusion, but we can restart with $X^{(0)}_{\text{NEW}} = K(X^{(0)}) \cap X^{(0)}$.

4. $K(X^{(0)})$ is not defined. In this case, we can bisect $X^{(0)}$ and process each half separately.

Note. $K(X^{(0)}) \subset \text{int}(X^{(0)}) \implies$ there exists a unique solution to $f(x) = 0$ in X only if f is continuous over X. However, it is useful in some types of interval systems to produce interval values even if part of the domain interval X lies outside the domain of the function. For example, if we have $y = x^{3/2} - 1/2$, and $X = \left[-\frac{1}{2}, \frac{1}{2}\right]$, then we might evaluate $\left[-\frac{1}{2}, \frac{1}{2}\right]^{3/2}$ by ignoring the part of X where $x^{3/2}$ is undefined, thus obtaining

$X^{3/2} \subseteq [0, 2^{-3/2}] \subseteq [0, 0.3536]$, and $x^{3/2} - 1/2 \subseteq [-0.5, -0.1464]$. A correct conclusion would be that if $y = x^{3/2} - 1$, then $y \in [-0.5, -0.1464]$. However, if we were looking at fixed-point iteration $x_{k+1} = g(x_k)$ with $g(x) = x^{3/2} - 1$, then the computation would seemingly prove that $g(X) \subseteq X$, leading to the false conclusion that g has a fixed point in X; indeed, the only solution to $g(x) = x$ lies in $[1.1913, 1.1917]$. The problem is that the fixed-point theorem "$g(X) \subseteq X \implies$ exists $x \in X$ such that $g(x) = x$" has "g is continuous over X" as one of its hypotheses. There had been discussion in the interval community about this for some time; in [194], Pryce proposed that systems implementing interval arithmetic perhaps allow evaluation over partial ranges, but raise a `discontinuous flag` if the function is not defined over part of the range.

Exercise 8.20. Issue the command `sqrt(infsup(-1,1))` in INTLAB. Explain what INTLAB returns. Is that a reasonable alternative for handling this case? Would this lead to incorrect conclusions concerning existence and uniqueness? □

There is a routine `in0` in INTLAB to check whether an interval (or whether the components of an interval vector or matrix) is contained in the interior of the components of another interval vector or matrix.

Exercise 8.21. Write a routine

```
[narrower_X,is_verified] = verified_existence_and_uniqueness(X,f)
```

that uses `Krawczyk_step` and `in0` in such a way that `is_verified` is set to 1 if there is a unique solution to $f(x) = 0$ in X and is set to 0 otherwise. Your routine `verified_existence_and_uniqueness` should set `narrower_X` to `X` if `is_verified` is set to 0 and should set `narrower_X` to the interval vector to which the Krawczyk method converges if `is_verified` is set to 1. Compare the output from your routine to the output from `verifynlss` for the problems you tried in Exercise 8.17. The inline documentation style in your routine should be the same as the style in the INTLAB examples in this text. (If you have `verified_existence_and_uniqueness.m` in MATLAB's search path, such as in MATLAB's working directory, and you type

```
help verified_existence_and_uniqueness
```

then MATLAB prints all comment lines, that is, all lines beginning with `%` up to the first blank line, to the command window.) □

Exercise 8.22. Use either `Krawczyk_step` or `verifynlss` to compute narrow bounds on the two solutions in Example 8.5.

8.4 Multivariate Interval Newton Methods

An alternative to the Krawczyk method can be developed more closely analogously to the univariate case. In particular, using the notation for f and F' from (8.11), (8.12), and Theorem 8.2, we have the following analogues to (8.5), (8.6), and (8.7). First, (8.5)

$$f(y) - f(x) = A(y - x) \tag{8.22}$$

corresponds to (8.6) for every x and y in the domain of f, where A is a matrix whose ith row is of the form[35]

$$A_{i,:} = \nabla^T f_i(c_i) = \left(\frac{\partial f_i}{\partial x_1}(c_i), \ldots, \frac{\partial f_i}{\partial x_n}(c_i) \right),$$

where $c_i \in \mathbb{R}^n$ is some point, usually not known explicitly, on the line between x and y. In general, A is not the Jacobian matrix $F'(c)$ for any particular $c \in \mathbb{R}^n$. However, if $F'(X)$ is an elementwise interval extension of the Jacobian matrix over some box X that contains both x and y, then, necessarily, $A \in F'(X)$. If we assume $f(x) = 0$ in (8.22), then

$$A(y - x) = f(y), \tag{8.23}$$

which is the analog of (8.5). Assuming for the time being that A is nonsingular, the analog to (8.6) is

$$x = y - A^{-1} f(y), \tag{8.24}$$

where A takes the place of $f'(s)$. In (8.6), we replaced $f'(s)$ by $F'(X)$ to obtain the interval Newton operator (8.8), but doing so in (8.24) would require us to define $F'(X)^{-1}$ for interval matrices $F'(X)$. Although that can be done clearly if not in a unique way (see [167]), to keep our explanation simple we elect instead to note that

$$x \in N(X) = y + V, \text{ where } V \text{ bounds the solution set to } F'(X)v = -F(y). \tag{8.25}$$

Thus, $N(X)$ here is the multidimensional analog to (8.8). The general interval Newton operator $N(X)$ has many of the same properties and can be used in much the same way as the Krawczyk operator $K(X)$. For instance, the following theorem, analogous to Theorem 8.3, is a special case of Theorem 5.1.7 in [167].

Theorem 8.4. Suppose f has continuous partial derivatives over X and $F'(X)$ is an elementwise interval extension to the Jacobian matrix of F over X. Suppose y is any point in X and suppose $N(X)$ is as in (8.25), where V is computed by any method for enclosing the solution set to the linear system $F'(X)v = -F(y)$. Then

$$N(X) \subset \text{int}(X) \implies \text{f has a unique solution in } X \text{ that is also in } N(X).$$

For example, the Krawczyk method (thought of as a method for solving linear systems, as on p. 91) or the interval Gauss–Seidel method (as on p. 96) can be used to bound the solution set to $F'(X)v = -F(y)$, and the four possibilities listed on p. 122 are valid with $N(X)$ replacing $K(X)$. In addition, if we use the interval Gauss–Seidel method or something similar, then a denominator $G_{i,i}$ in (7.10) could contain zero, and we have a fifth possibility:

5. $N(X^{(0)}) \cap X^{(0)}$ is the union of two boxes; we can process each of these two boxes separately.

[35]This "colon" notation is what MATLAB uses to access rows and columns of matrices. That is, `A(i,:)` produces the ith row of the matrix A (as a row vector), while `A(:,j)` produces the jth column of the matrix A, as a column vector. This notation has been adopted in much literature on numerical linear algebra, most notably in the classic advanced text [50].

For example, the following nonlinear_Gauss_Seidel_image function can be used in place of the Krawczyk_step function on p. 118:

```
function [NX,error_occurred] = nonlinear_Gauss_Seidel_image(X,y,f)
% [NX,error_occurred] = nonlinear_Gauss_Seidel_image(X,y,f)
% returns the image of X from a single sweep of the nonlinear
% Gauss--Seidel method, with base point y, where
% the function is programmed in the function whose name is in
% the string f.  If error_occurred = 1, no image NX is returned;
% otherwise, error_occurred = 0 is returned.
% There is no need to program the Jacobian matrix,
% since the "gradient" automatic differentiation toolbox,
% distributed with INTLAB, is used.  It is the user's
% responsibility to ensure that X is an interval column vector,
% that y is a non-interval column vector contained in X, and
% that f is the desired function in an "m" file that returns
% a column vector.


% First compute f(y) using interval arithmetic to bound
% roundoff error --
n = length(X);
iy = midrad(y,0);
fy = feval(f,iy);

% Now compute F'(X) and the preconditioning matrix Y --
Xg = gradientinit(X);
FXg = feval(f,Xg);

% Compute the initial V --
V = X-y;
% Now, do the Gauss--Seidel sweep to find V --
[new_V,is_empty,error_occurred] = Gauss_Seidel_image(FXg.dx, -fy, V);

NX - y+new_V;
```

Example 8.6. We will redo Example 8.4 using the interval Gauss–Seidel method as implemented in nonlinear_Gauss_Seidel_image to bound the solution to the system of linear equations. We obtain

```
>> X = [infsup(.5,.8); infsup(.6,.9)]
intval X =
[ 0.5000 , 0.8000 ]
[ 0.5999 , 0.9000 ]
>> y=mid(X)
y =
    0.6500
    0.7500
>> [NX, error_occurred] =
           nonlinear_Gauss_Seidel_image(X,y,'example8p4')
```

```
intval NX =
[ 0.5687 , 0.6588 ]
[ 0.7702 , 0.8130 ]
error_occurred = 0
 .
 .
 .
intval X =
[ 0.6172 , 0.6179 ]
[ 0.7848 , 0.7853 ]
>> [NX, error_occurred] =
           nonlinear_Gauss_Seidel_image(X,y,'example8p4')
intval NX =
[ 0.6175 , 0.6176 ]
[ 0.7850 , 0.7851 ]
error_occurred = 0
```

Observe that the image after the first iteration is contained in the corresponding image after the first iteration of the Krawczyk method. This is true in general; the result occurs in [167] (and possibly earlier). For this example, this is due to intrinsic mathematical properties of the Krawczyk method and the Gauss–Seidel method, not to the fact that the matrix Y in (8.18) is only an approximation. To see this, we can redo the computation with the loose approximation (8.18):

```
>> X = [infsup(.5,.8); infsup(.6,.9)]
intval X =
[ 0.5000 , 0.8000 ]
[ 0.5999 , 0.9000 ]
>> y=mid(X)
y =
    0.6500
    0.7500
>> fprimex = [2*X(1) 2*X(2); 1 -2*X(2)]
intval fprimex =
[ 1.0000 , 1.6000] [ 1.1999, 1.8000 ]
[ 1.0000 , 1.0000] [ -1.8000, -1.1999 ]
>> fy = [y(1)^2 + y(2)^2-1;y(1)-y(2)^2]
fy =
   -0.0150
    0.0875
>> y = midrad(y,0)
intval y =
[ 0.6499 , 0.6501 ]
[ 0.7499 , 0.7501 ]
>> V = X-y
intval V =
[ -0.1501 , 0.1500 ]
[ -0.1501 , 0.1500 ]
>> Y = [.43 .43; .29 -.37]
Y =
    0.4300    0.4300
    0.2900   -0.3700
```

```
>> [new_V_1, V12, there_are_2, num, denom] ...
 = gauss_seidel_step(1,Y(1,:),fprimex, -fy, V)
intval new_V_1 = [ -0.0813 , 0.0088 ]
intval V12 = [ -0.1501 , 0.1500 ]
there_are_2 = 0
intval num = [ -0.0699 , 0.0076 ]
intval denom = [ 0.8599 , 1.1181 ]
>> new_X(1) = y(1)+new_V_1
intval new_X = [ 0.5687 , 0.6588 ]
```

□

Note We have stated the multidimensional interval Newton method somewhat more generally than the univariate one. In the univariate case, we fixed the base point y to be the midpoint of X, but here, we left y arbitrary other than $y \in X$. In fact, the only requirement for the univariate case is also $y \in X$, and there may be some advantages to choosing y other than $m(X)$.

8.5 Concluding Remarks

See Hansen and Walster [57] (especially their Chapters 9 and 11) for a thorough discussion of interval Newton methods for nonlinear equations and nonlinear systems of equations. A treatment of the theory underlying the Krawczyk operator, the interval Gauss–Seidel operator, and other aspects of interval techniques for linear and nonlinear systems appears in [167].

It is possible to use a slope matrix (see p. 72) instead of a componentwise extension to the Jacobian matrix in the interval Newton method. In [257], Shen and Wolfe study slope enclosures and show that, in using them in the Krawczyk method, the method has higher existence-proving power than the Newton–Kantorovich theorem.[36] However, if slopes are used in interval Newton methods, uniqueness is lost. In [219], Rump discusses this and explains an efficient two-stage method whereby slopes can be used to prove both existence and uniqueness.

A number of experts in interval computations have produced software that automatically finds all solutions, with mathematical rigor, to systems of nonlinear equations. Some of our work, publicly available, is GLOBSOL. GLOBSOL implements a branch-and-bound method in combination with computational existence and uniqueness based on a multivariate interval Newton method to find all global optimizers of unconstrained or constrained optimization problems or (with configuration variable `NONLINEAR_SYSTEM` set to `.TRUE.`) all solutions to nonlinear systems of equations within a particular box. A brief description of the history and capabilities of GLOBSOL can be found in [104], while an in-depth description of some of the techniques in GLOBSOL, along with details concerning a very early version of GLOBSOL (which was called INTOPT_90), can be found in [97]. The system of equations or optimization problem is provided to GLOBSOL in the form of a Fortran 90 program. A self-contained, easily installed version of GLOBSOL (distributed with a copy of the GNU Fortran 95 compiler) for MS Windows is available.

[36] One place where this theorem is stated is [188, p. 421]. Interestingly, the hypotheses of the Newton–Kantorovich theorem can be verified using interval arithmetic.

Interval versions of Newton's method in Banach spaces are presented in Moore and Cloud [156].

While the previous two chapters dealt with linear and nonlinear algebra, the next chapter will use interval evaluations to bound error terms, for numerical integration with verified error bounds.

Chapter 9

Integration of Interval Functions

In this chapter we introduce the interval integral

$$\int_{[a,b]} F(t)\,dt$$

and use it to compute enclosures for the definite integrals of real-valued functions. We also revisit the subject of *automatic differentiation*, whose origins date to the appearance of modern computers [92], which appeared in the context of interval analysis in [157], and which is an extensive subject in its own right [21, 54, 200]. Finally, we outline some relationships between interval integration and the other, more familiar types of integration (Riemann, Lebesgue, etc.).

9.1 Definition and Properties of the Integral

Let us begin with a real-valued continuous function f. We typically denote the ordinary definite integral

$$\int_a^b f(t)\,dt \qquad \text{by either} \qquad \int_{[a,b]} f(t)\,dt \quad \text{or} \quad \int_X f(t)\,dt,$$

where $X = [a, b]$.

Lemma 9.1. If f is continuous in $X = [a, b]$, then

$$\int_X f(t)\,dt \in f(X)w(X). \tag{9.1}$$

Proof. According to the mean value theorem for integrals, we can write

$$\int_{[a,b]} f(t)\,dt = f(s)(b-a) \text{ for some } s \in [a,b]. \tag{9.2}$$

But $f(s) \in f(X)$, so the lemma follows. □

Theorem 9.1. Suppose $f \in \mathrm{FC}_1(X_0)$ and F is an inclusion isotonic, Lipschitz, interval extension of f with $F(X)$ defined for all $X \subseteq X_0$. Let N be a positive integer and subdivide $[a, b] \subseteq X_0$ into N subintervals $X_1, \ldots, X_N$ so that

$$a \leq \underline{X_1} < \overline{X_1} = \underline{X_2} < \overline{X_2} < \cdots < \overline{X_N} = b.$$

Then

$$\int_{[a,b]} f(t)\,dt \in \sum_{i=1}^{N} F(X_i)w(X_i). \tag{9.3}$$

Furthermore, there is a constant L, independent of both N and the mode of subdivision, such that

$$w\left(\sum_{i=1}^{N} F(X_i)w(X_i)\right) \leq L \sum_{i=1}^{N} w(X_i)^2. \tag{9.4}$$

Proof. Because $\bar{f}(X) \subseteq F(X)$, we have

$$\int_X f(t)\,dt \in F(X)w(X) \text{ for all } X \subseteq X_0 \tag{9.5}$$

by Lemma 9.1. Therefore,

$$\int_{X_i} f(t)\,dt \in F(X_i)w(X_i) \qquad (i = 1, \ldots, N).$$

This and the additivity property

$$\int_{[a,b]} f(t)\,dt = \sum_{i=1}^{N} \int_{X_i} f(t)\,dt \tag{9.6}$$

yield (9.3). Finally, by Definition 6.1, there is a constant L such that $w(F(X)) \leq Lw(X)$ for all $X \subseteq [a, b] \subseteq X_0$. We have

$$w\left(\sum_{i=1}^{N} F(X_i)w(X_i)\right) = \sum_{i=1}^{N} w(F(X_i))w(X_i) \leq L \sum_{i=1}^{N} w(X_i)^2$$

as desired. ☐

For a uniform subdivision of $[a, b]$ with

$$w(X_i) = (b - a)/N \qquad (i = 1, \ldots, N), \tag{9.7}$$

define

$$S_N = S_N(F; [a, b]) = \sum_{i=1}^{N} F(X_i)(b - a)/N. \tag{9.8}$$

We have the following theorem.

Theorem 9.2.

$$\int_{[a,b]} f(t)\,dt = \bigcap_{N=1}^{\infty} S_N(F;[a,b]) = \lim_{N\to\infty} S_N(F;[a,b]). \tag{9.9}$$

Proof. By (9.7), inequality (9.4) becomes

$$w(S_N) \leq L(b-a)^2/N. \tag{9.10}$$

Therefore, $w(S_N) \to 0$ as $N \to \infty$, with the value of the integral lying in every S_N by (9.3). □

It follows from Lemma 6.5 that the sequence of intervals defined by

$$Y_1 = S_1, \qquad Y_{k+1} = S_{k+1} \cap Y_k \quad (k = 1, 2, \ldots)$$

is nested and converges to the value of the integral. We have finite convergence if IA is used.

We can use Theorem 9.2 to define the integral of an *interval-valued* function of a real variable t. However, we first note that, while $F(t)$ must be real-valued for a Lipschitz interval extension F, this is not true for all useful inclusion isotonic interval functions. The interval polynomial enclosures discussed later have interval number coefficients. In such cases, the width of $F(t)$ may not be zero for real t.

Suppose that an interval-valued function with values $F(X)$ is continuous and inclusion isotonic for $X \subseteq X_0$. If $[a,b] \subseteq X_0$, the sums $S_N(F;[a,b])$ given in (9.8) are well defined. For real arguments t (degenerate intervals $[t,t]$), the values of $F(t)$ may be real numbers or intervals.

Definition 9.1. We define the interval integral

$$\int_{[a,b]} F(t)\,dt = \bigcap_{N=1}^{\infty} S_N(F;[a,b]). \tag{9.11}$$

It follows from the continuity of F that there are two continuous real-valued functions $\underline{F}$ and $\overline{F}$ such that, for real t,

$$F(t) = \left[\underline{F}(t), \overline{F}(t)\right].$$

Moreover, the integral defined by (9.11) is equivalent to

$$\int_{[a,b]} F(t)\,dt = \left[\int_{[a,b]} \underline{F}(t)\,dt\,,\ \int_{[a,b]} \overline{F}(t)\,dt\right]. \tag{9.12}$$

Example 9.1. Suppose $0 < a < b$ and let $F(t) = A\,t^k$, where $A = \left[\underline{A}, \overline{A}\right]$ and k is a positive integer. By (9.12) we have

$$\begin{aligned}
\int_{[a,b]} F(t)\,dt &= \left[\int_{[a,b]} \underline{A}\,t^k\,dt\,,\ \int_{[a,b]} \overline{A}\,t^k\,dt\right] \\
&= \left[\underline{A}\,\frac{b^{k+1}-a^{k+1}}{k+1}\,,\ \overline{A}\,\frac{b^{k+1}-a^{k+1}}{k+1}\right] \\
&= A\,\frac{b^{k+1}-a^{k+1}}{k+1}.
\end{aligned}$$

□

Exercise 9.1. Rework Example 9.1 for the case where $a < 0 < b$ and k is an odd positive integer. □

We will discuss the integration of interval polynomials in the next section.

Computation with INTLAB

The computation in (9.8) can be done with a routine that is similar in structure to the routine `refinement` on p. 56, except that $w(X_i)$ needs to be computed with interval arithmetic to take roundoff error into account. Additionally, we use the internal MATLAB helper routine `fcnchk` that allows us to pass either a string defining the function or a string giving the m-file defining the function. We obtain

```
function S_N = integral_sum(X,f,N)
% Y = S_N = integral_sum(X,f,N)
% computes the sum defined in (9.8) of the text.

strfun = fcnchk(f);

% First form the N subintervals --
h = (midrad(sup(X),0)-midrad(inf(X),0))/midrad(N,0);
% h represents w(X_i) as in (9.7) of the text.
% The computations above and below should be done with intervals
% so roundoff error is taken into account.
xi = midrad(inf(X),0);
x1 = xi;
for i=1:N
    xip1 = x1 + i*h;
    Xs(i) = infsup(inf(xi),sup(xip1));
    xi = xip1;
end

% Now compute the sum --
S_N = feval(strfun,Xs(1));
if N > 1
    for i=2:N
        S_N = S_N + feval(strfun,Xs(i));
    end
end
S_N = S_N * h;
```

For example, to compute bounds on $\int_{[0,1]} x^2 dx$, we might have the following dialogue in the MATLAB command window:

```
>> value = integral_sum(infsup(0,1),'x^2',5)
intval value = [ 0.2399 , 0.4401 ]
>> value = integral_sum(infsup(0,1),'x^2',50)
intval value = [ 0.3233 , 0.3435 ]
>> value = integral_sum(infsup(0,1),'x^2',500)
intval value = [ 0.3323 , 0.3344 ]
```

```
>> value = integral_sum(infsup(0,1),'x^2',5000)
intval value = [ 0.3332 , 0.3335 ]
```

Just as Riemann sums are not the most efficient or accurate way to approximately evaluate integrals using floating point arithmetic, (9.8) is not the most efficient or sharpest way to obtain bounds on an interval integral. For instance, the call above to `integral_sum` with `N=5000` took a noticeable amount of time, in part because that execution of explicitly programmed loops in MATLAB is much slower than in compiled languages. However, efficient implementations of more sophisticated methods to get sharp bounds for the interval integral exist. We will explain some of the more advanced techniques later in this chapter.

Exercise 9.2. Use `integral_sum` with various `N` to compute bounds on the following integrals:

(a) the elliptic integral of the first kind $F(k,\phi) = \int_0^{\phi}(1-k^2\sin^2 t)^{-1/2}\,dt$ with parameters $\phi = 2\pi$ and $k = 0.5$;

(b) the Fresnel sine integral $f(x) = \int_0^x \sin(t^2)\,dt$ at $x = 100$. □

Exercise 9.3. If you have Aberth's RANGE software, repeat Exercise 9.2 using that software.[37] □

Inclusion Property

We end this section by stating one more property of the interval integral:

$$\text{If } F(t) \subseteq G(t) \text{ for all } t \in [a,b], \text{ then } \int_{[a,b]} F(t)\,dt \subseteq \int_{[a,b]} G(t)\,dt. \tag{9.13}$$

In other words, interval integration preserves inclusion.

Exercise 9.4. Prove (9.13). □

9.2 Integration of Polynomials

Exercise 9.1 shows that care must be taken when integrating power-type functions having interval coefficients. Here we will summarize the results for an interval polynomial of the form

$$P(t) = A_0 + A_1 t + \cdots + A_q t^q, \tag{9.14}$$

where t is real and $A_i = [\underline{A}_i, \overline{A}_i]$ for $i = 0, 1, \ldots, q$. There are two cases:

Case 1. If a and b have the same sign, then

$$\int_{[a,b]} P(t)\,dt = A_0(b-a) + \cdots + A_q(b^{q+1} - a^{q+1})/(q+1).$$

[37] See the web page for this book, www.siam.org/books/ot110.

Case 2. If $a < 0 < b$, then

$$\int_{[a,b]} P(t)\,dt = T_0 + T_1 + \cdots + T_q,$$

where

$$T_i = \begin{cases} A_i(b^{i+1} - a^{i+1})/(i+1), & i \text{ even}, \\ \left[(\underline{A}_i b^{i+1} - \overline{A} a^{i+1})/(i+1),\ (\overline{A}_i b^{i+1} - \underline{A} a^{i+1})/(i+1)\right], & i \text{ odd}. \end{cases}$$

Note that for "real" (degenerate interval) coefficients A_i, with $\underline{A}_i = \overline{A}_i$, both cases reduce to the usual formal integration of polynomials.

INTLAB Implementation

The following MATLAB function implements the above formulas:

```
function I = interval_poly_integral(X,A)
% I = interval_poly_integral(X,A) returns the interval integral
% over the interval X of the polynomial with interval coefficients
% given by A(1) + A(2) x + ... + A(q+1) x^q, using the
% formulas given in Section 8.2 of the text by Moore, Kearfott,
% and Cloud.
q = length(A)-1;
a = inf(X);
b = sup(X);
ia = midrad(a,0);
ib = midrad(b,0);
if ( (a >=0) & (b >=0) ) | ( (a <=0) & (b <=0) )
    I = midrad(0,0);
    for i=1:q+1;
        I = I + A(i)*(ib^i - ia^i)/midrad(i,0);
    end
else
    I = midrad(0,0);
    for i=1:q+1;
        if mod(i,2)
            T_i = A(i)*(ib^i-ia^i)/midrad(i,0);
        else
            lb = ( midrad(inf(A(i)),0) * ib^i...
                  -midrad(sup(A(i)),0) * ia^i )...
                  / midrad(i,0)
            ub = ( midrad(sup(A(i)),0) * ib^i ...
                  -midrad(inf(A(i)),0) * ia^i ) ...
                  / midrad(i,0)
            T_i = infsup(inf(lb),sup(ub));
        end
        I = I + T_i;
    end
end
```

Exercise 9.5. Try `interval_poly_integral` to compute

$$\int_{[-1,1]} \left\{[1,2] + [-3,4]x + [-0.5,0.5]x^2\right\} dx.$$

Check the result by hand. □

9.3 Polynomial Enclosure and Automatic Differentiation

The main goal of this chapter is to discuss the computation of enclosures for the values of definite integrals such as

$$\int_{[a,b]} f(t)\,dt.$$

Broadly speaking, the approach is to (1) enclose $f(t)$ with a suitable interval function $G(t)$ and then (2) integrate $G(t)$. We carry out this procedure in section 9.4. The present section is devoted to step (1)—the definition and efficient computation of useful function enclosures.

Definition 9.2. Let T_0 be an interval and x a real-valued function of the real variable t, with $x(t)$ defined for all $t \in T_0$. An *interval enclosure* of x is an inclusion isotonic interval-valued function X of an interval variable T, with $X(T)$ defined for all $T \subseteq T_0$, having the property that

$$x(t) \in X(t) \text{ for all } t \in T_0. \tag{9.15}$$

Hence, $x(T) \subseteq X(T)$ for all $T \subseteq T_0$.

Example 9.2. Suppose $x(t)$ satisfies $a \le x(t) \le b$ for all $t \in T_0$. Then the constant interval function $X(T) \equiv [a,b]$ is an interval enclosure of x. □

Note that $X(t)$ in (9.15) is not necessarily real; it may be a nondegenerate interval for each value of t. If $X(t) = x(t)$ for all $t \in T_0$, then X is an inclusion isotonic interval extension of x. This latter case will hold if $X(t)$ takes the form of a polynomial in t with interval coefficients, i.e., the form (9.14).

Definition 9.3. If $X(t)$ in Definition 9.2 takes the form (9.14), it is called an *interval polynomial enclosure* of x.

Example 9.3. Suppose $x(t)$ is continuously differentiable on T_0 and that $X'(T)$ is an inclusion isotonic interval extension of $x'(t)$. By the mean value theorem we have, for any $t_0 \in T_0$,

$$x(t) = x(t_0) + x'(s)(t - t_0) \text{ for some } s \text{ between } t \text{ and } t_0. \tag{9.16}$$

It follows that $X(T) = x(t_0) + X'(T_0)(T - t_0)$ is an interval enclosure of x. The interval-valued function $X(t)$ of the real variable t is an interval polynomial enclosure of x. We have

$$x(t) \in X(t) = x(t_0) + X'(T_0)(t - t_0) \text{ for all } t \in T_0. \tag{9.17}$$

□

We can extend the idea of (9.16) by working with Taylor expansions, for which we introduce some notation. Suppose $x(t)$ is analytic in t in some neighborhood of t_0. We define

$$(x)_0 = x(t_0),$$
$$(x)_k = \frac{1}{k!}\frac{d^k x}{dt^k}(t_0) \qquad (k = 1, 2, \ldots). \tag{9.18}$$

That is, $(x)_k$ is the kth Taylor coefficient in the expansion of $x(t)$ about $t = t_0$:

$$x(t) = \sum_{k=0}^{\infty}(x)_k(t - t_0)^k. \tag{9.19}$$

For the finite Taylor expansion with Lagrange form of the remainder, we have

$$x(t) = \sum_{k=0}^{N-1}(x)_k(t - t_0)^k + (x)_N(s)(t - t_0)^N \quad \text{for some } s \in [t_0, t]. \tag{9.20}$$

If R_N is an interval enclosure of $x_N(s)$ so that

$$(x)_N(s) \in R_N([t_0, t]) \text{ for all } s \in [t_0, t],$$

for t within the radius of convergence of (9.19), we have

$$x(t) \in \sum_{k=0}^{N-1}(x)_k(t - t_0)^k + R_N([t_0, t])(t - t_0)^N. \tag{9.21}$$

The right-hand side is an enclosure for the Taylor series with remainder (9.20).

Example 9.4. If $x(t)$ is analytic for $t \in T_0$, the interval polynomial

$$X_N(t) = \sum_{k=0}^{N-1}(x)_k(t - t_0)^k + R_N(T_0)(t - t_0)^N \tag{9.22}$$

is an enclosure of x. The values of $X_N(T)$ for nondegenerate intervals $T \subseteq T_0$ are also defined, and we have

$$x(t) \in X_N(t) \subseteq X_N(T) = \sum_{k=0}^{N-1}(x)_k(T - t_0)^k + R_N(T_0)(T - t_0)^N \tag{9.23}$$

for all $t \in T \subseteq T_0$. As a specific application of (9.22), let $x(t) = e^t$ and take $t_0 = 0$, $T_0 = [0, 1]$, $N = 2$. We have

$$e^t \in 1 + t + \tfrac{1}{2}[1, e]\,t^2 = \left[1 + t + \tfrac{1}{2}t^2,\ 1 + t + \tfrac{e}{2}t^2\right] \quad \text{for all } t \in [0, 1].$$

For a finite representation, which is still an interval polynomial enclosure, we can take $X(t) = 1 + t + [0.5, 1.359141]\,t^2$. For a sharper enclosure, we can take $N = 10$ in (9.22) to obtain

$$e^t \in X_{10}(t) = 1 + t + \tfrac{1}{2}t^2 + \cdots + \tfrac{1}{9!}t^9 + \tfrac{1}{10!}[1, e]\,t^{10}.$$

Only the last coefficient is a nondegenerate interval. It is contained in the finitely represented interval $[0.27 \cdot 10^{-6}, 0.75 \cdot 10^{-6}]$. We can enclose the real coefficients in narrow intervals to obtain a finitely represented interval polynomial enclosure $X(t)$ which contains $X_{10}(t)$ for every $t \in [0, 1]$. Of course, $X(t)$ also contains e^t for every $t \in [0, 1]$. □

Note Use of Taylor polynomials in this way is a heuristic that often, but not always, reduces overestimation and excess width as we take polynomials and error terms of increasingly higher order. For example, if high-order Taylor polynomials centered on $t = 0$ were used to approximate e^t for $|t|$ large and t negative, then the interval widths would increase as higher-order polynomials were taken. This phenomenon is analogous to the cancelation error we would observe, were we to approximate such e^t with a high-order Taylor polynomial with floating point arithmetic.

It remains to discuss how the Taylor coefficients $(x)_k$ can be evaluated efficiently. An *inefficient* approach would be to use (9.18) to generate formulas for the $(x)_k$ in terms of t and then evaluate them at t_0. Indeed, the expressions for the successive derivatives of practical functions x typically become so involved that things can seem hopeless for all but the smallest values of k. Fortunately, it is *not* necessary to generate explicit formulas for the $(x)_k$ to obtain the *values* $(x)_k(t_0)$. The latter can be computed numerically using the techniques of *automatic differentiation*.

The approach amounts to differentiating computer subroutines (or function subprograms). The result of differentiating a program is another program. When executed, the derived program produces a value of the derivative of the function defined by the first program. In fact, we can write a general-purpose program which, when applied to a function defined by another program, will produce a Kth derivative program for general K for that function. If we supply it with an integer K and argument values, it will produce values of the derivatives (or Taylor coefficients) up to order K of the given function at the given argument values. This can be carried out in machine arithmetic or in IA. Partial derivatives can be obtained in a similar way. Early references on this are [147, 148, 157]; a somewhat more recent one is [200], while a more recent, comprehensive overview in technical terms is [53].

We will now show how to compute the real and interval values $(x)_k$ and $R_N([t_0, t])$. The technique will apply to almost any quantity dependent on x that is computed with a computer program. From (9.18) we find that $x'(t_0) = (x)_1$, so

$$(x)_k = \frac{1}{k}((x)_1)_{k-1}. \tag{9.24}$$

This relation is important, because if we have a function $x(t)$ defined by a differential equation, for instance,

$$x'(t) = g(t, x(t)) \text{ with a given } x(t_0),$$

then we can use

$$(x)_k = \frac{1}{k}(g)_{k-1}$$

recursively to compute the Taylor coefficients for $x(t)$ about t_0—provided, of course, that we can handle $g(t, x(t))$. Similarly,

$$(x)_k = \frac{2}{k(k-1)}((x)_2)_{k-2}.$$

As far as the arithmetic operations are concerned, if u and v are analytic functions of t in some neighborhood of $t = t_0$, then

$$\left.\begin{aligned}
(u+v)_k &= (u)_k + (v)_k, \\
(u-v)_k &= (u)_k - (v)_k, \\
(uv)_k &= \sum_{j=0}^{k} (u)_j (v)_{k-j}, \\
\left(\frac{u}{v}\right)_k &= \left(\frac{1}{(v)_0}\right)\left\{(u)_k - \sum_{j=1}^{k} (v)_j \left(\frac{u}{v}\right)_{k-j}\right\}.
\end{aligned}\right\} \quad (9.25)$$

Exercise 9.6. Derive a few of these formulas. □

All the commonly used "elementary" functions satisfy rational differential equations. Any second-order differential equations that may occur can be rewritten as pairs of first-order equations. With this and (9.24), we can derive *recursion relations* for the kth Taylor coefficients of elementary functions of an arbitrary analytic function $u(t)$. For example,[38]

$$\left.\begin{aligned}
(u^a)_k &= \left(\frac{1}{u}\right)\sum_{j=0}^{k-1}\left(a - \frac{j(a+1)}{k}\right)(u)_{k-j}(u^a)_j, \\
(e^u)_k &= \sum_{j=0}^{k-1}\left(1 - \frac{j}{k}\right)(e^u)_j (u)_{k-j}, \\
(\ln u)_k &= \left(\frac{1}{u}\right)\left((u)_k - \sum_{j=1}^{k-1}\left(1 - \frac{j}{k}\right)(u)_j (\ln u)_{k-j}\right), \\
(\sin u)_k &= \left(\frac{1}{k}\right)\sum_{j=0}^{k-1}(j+1)(\cos u)_{k-1-j}(u)_{j+1}, \\
(\cos u)_k &= -\left(\frac{1}{k}\right)\sum_{j=0}^{k-1}(j+1)(\sin u)_{k-1-j}(u)_{j+1}.
\end{aligned}\right\} \quad (9.26)$$

Similar relations are available for the hyperbolic functions, the Bessel functions, and various other functions commonly used in applied mathematics.

Exercise 9.7. Show the following:

(a) If $u(t) =$ constant, then $(uv)_k = u(v)_k$.

[38] In the relation for $(\ln u)_k$, the sum is deleted if $k = 1$.

(b) When $u(t) = t$, we have

$$(e^t)_k = \frac{1}{k}(e^t)_{k-1} \quad \text{and the pair} \quad \begin{cases} (\sin t)_k = \frac{1}{k}(\cos t)_{k-1}, \\ (\cos t)_k = -\frac{1}{k}(\sin t)_{k-1}. \end{cases}$$

The sine and cosine formulas should be computed together, even if only one is needed directly. This will be illustrated in Example 9.6. □

A general procedure for computing the Taylor coefficients of functions $x(t)$ can be programmed to operate as follows:

1. Represent $x(t)$ (or $x'(t)$ or $x''(t)$ $\cdots$) by a finite list of binary or unary operations (e.g., $T_3 = T_1 + T_2$, $T_4 = e^{T_3}$, etc.).

2. On a line-by-line basis, generate subprograms for Taylor coefficients for each item in the list, using the recursion relation appropriate for the operation in that item.

3. Organize the subprograms and the data handling so that the derived program will accept initial values for N, t_0, $x(t_0)$ (and $x'(t_0)$, $x''(t_0)$, ..., if required); the derived program will evaluate and store, in order, the first Taylor coefficients of each item in the list, then the second Taylor coefficient of each item in the list (which may require some of the stored values of the first coefficients and initial data), etc., until the entire array of coefficients has been computed; the process can be carried out either in real computer arithmetic or in IA.

4. The list in step 1 can be generated from a subprogram description of $x(t)$.

Example 9.5. Consider the function $x(t)$ defined (when t_0 and $x(t_0)$ are given) by the differential equation

$$x'(t) = x^2 + t^2. \tag{9.27}$$

For simplicity (to use the formula in (9.25) for multiplication, rather than the first formula in (9.26)), we write x^2 as $x \cdot x$ and t^2 as $t \cdot t$. (We would prefer (9.26) for sharpness in interval evaluations.) We have

$$\begin{aligned} (T_1)_0 &= t \cdot t, \\ (T_2)_0 &= (x)_0 \cdot (x)_0, \\ (x)_1 &= T_1 + T_2. \end{aligned} \tag{9.28}$$

Applying the third recursion in (9.25), we obtain

$$\left.\begin{aligned} (T_1)_k &= \sum_{j=0}^{k} (t)_j (t)_{k-j}, \\ (T_2)_k &= \sum_{j=0}^{k} (x)_j (x)_{k-j}, \\ (x)_{k+1} &= ((T_1)_k + (T_2)_k)/(k+1). \end{aligned}\right\} \tag{9.29}$$

Since $(t)_1 = 1$ and $(t)_j = 0$ for $j > 1$, the straightforward application of the recursion relations (9.29) could be modified to recognize that $(T_1)_k = 0$ when $k \geq 3$. Even without any such simplification, the relations (9.29) suffice for the numerical evaluation of any number of Taylor coefficients $(x)_k$, $k = 1, 2, \ldots$, for the function $x(t)$. The total number of arithmetic operations required to find $(x)_k$ for $k = 1, \ldots, N$ not taking into account simplifications possible because of zero terms is (from (9.28) and (9.29)) $N^2 + N - 1$ additions, $N^2 + N$ multiplications, and $N - 1$ divisions. □

Exercise 9.8. Repeat Example 9.29 using $T_1 = t^2$ and $T_2 = x^2$. Assuming $t_0 = 1$ and $x(t_0) = 0$, use the recursions you so obtain to compute the first five Taylor coefficients for $x(t)$ expanded about $t = 1$.

The number of arithmetic operations to compute the kth Taylor coefficient of any function expressible as a finite combination of rational, elementary, and composition functions grows at most linearly with k once the coefficients of lower orders have been obtained and stored by the process described. Thus, the total number of arithmetic operations to obtain the set of Taylor coefficients of orders $1, \ldots, N$ grows no faster than cN^2 for some constant c independent of N and depending only on the particular function in question.

Example 9.6. Consider

$$x(t) = e^{-\sin t} \ln(1 + t). \tag{9.30}$$

We can represent $x(t)$ by the list

$$\begin{aligned} T_1 &= \sin t, \\ T_2 &= \ln(1 + t), \\ T_3 &= e^{-T_1}, \\ T_4 &= \cos t, \\ x(t) &= T_2 \cdot T_3. \end{aligned} \tag{9.31}$$

Applying the appropriate relations from (9.25) and (9.26) on a line-by-line basis, we obtain, using $(t)_0 = t$, $(t)_1 = 1$, and $(t)_j = 0$ for $j > 1$,

$$\begin{aligned} (T_1)_k &= \frac{1}{k}(T_4)_{k-1}, \\ (T_2)_k &= \frac{1}{1+t}\left(a_k - \left(1 - \frac{1}{k}\right)(T_2)_{k-1}\right), \quad a_k = \begin{cases} 1, & k = 1, \\ 0, & k > 1, \end{cases} \\ (T_3)_k &= \sum_{j=0}^{k-1}\left(1 - \frac{j}{k}\right)(T_3)_j(-T_1)_{k-j}, \\ (T_4)_k &= -\frac{1}{k}(T_1)_{k-1}, \\ (x)_k &= \sum_{j=0}^{k}(T_2)_j(T_3)_{k-j}. \end{aligned} \tag{9.32}$$

We compute and store the following two-dimensional array of quantities proceeding downward through the elements of a given column before continuing with the first element of the next column to the right:

$$\begin{array}{cccc} T_1 & (T_1)_1 & (T_1)_2 & \cdots \\ T_2 & (T_2)_1 & (T_2)_2 & \cdots \\ T_2 & (T_2)_1 & (T_2)_2 & \cdots \\ T_4 & (T_4)_1 & (T_4)_2 & \cdots \\ x & (x)_1 & (x)_2 & \cdots \end{array}$$

Again the total arithmetic computation required to find the Taylor coefficients of orders $1, \ldots, N$ for $x(t)$ defined by (9.30) at any value of t is of order N^2 (about $\frac{3}{2}N^2$ additions and multiplications). Note that the functions $\sin t$, $\cos t$, $\ln(1+t)$, $e^{-\sin t}$ need to be evaluated only *once* each, at the outset of the process.

We can bound the remainder term in (9.20) for this $x(t)$ by using interval extensions of $\sin t$, $\cos t$, $\ln(1+t)$, and e^{-T_1}. Carrying out evaluations of those interval extensions for the interval $[t_0, t]$ and following this with evaluations of the Taylor coefficients, using (9.32) in interval arithmetic, we can obtain $R_N([t_0, t])$ in (9.21). □

Notes

1. Martin Berz, Kyoko Makino, and collaborators have provided an implementation of computation of multivariate Taylor coefficients that is highly successful in applications; see the COSY web page.

2. In many instances, it is more efficient to arrange the computations in automatic differentiation in a different order than has been presented here. This is particularly so if just the function, gradient, and, perhaps, directional second-order derivatives are needed. The computational scheme presented here is commonly called the *forward mode*. To compute the value of gradient of a function of n variables, the forward mode typically requires n times the amount of work required to compute the value of the function itself. An alternative is the *backward mode*, in which the intermediate results in the computation are first computed and stored, then a system is solved to obtain the values. In contrast to the forward mode, computing the values of the function and the gradient using the backward mode requires only a constant multiple (around 5) times the amount of work required to compute just the value of the function but requires storage that is proportional to the number of operations required to compute the function. The backward mode was perhaps first introduced by Speelpenning [235] and has since been strongly advocated by Griewank [53], Christianson [36], and others. The backward mode is explained in elementary terms in texts such as [3]. A comprehensive monograph including details of efficient implementation of both the forward mode and backward mode is [53].

9.4 Computing Enclosures for Integrals

It follows from (9.13) that if G is an interval polynomial enclosure of f, then

$$\int_{[a,b]} f(t)\,dt \in \int_{[a,b]} G(t)\,dt. \tag{9.33}$$

More generally, (9.33) holds if f is real-valued and G is interval-valued, continuous, and inclusion isotonic, with $f(t) \in G(t)$ for all $t \in [a, b]$.

Example 9.7. Consider $f(t) = 1/t$ and put $F(X) = 1/X$. From (9.5) we have

$$\int_{[1,2]} (1/t)\, dt \in (1/[1,2]) = \left[\tfrac{1}{2}, 1\right]. \tag{9.34}$$

From (9.8), we have

$$\begin{aligned} S_N &= \sum_{i=1}^{N} (1/(1 + [i-1, i]/N))/N \\ &= \sum_{i=1}^{N} [1/(N+i), 1/(N+i-1)]. \end{aligned}$$

Using three-digit IA we obtain

$$\begin{aligned} S_1^* &= [0.5, 1.0], \\ S_2^* &= [0.583, 0.834], \\ S_3^* &= [0.617, 0.783], \\ &\vdots \\ S_{10}^* &= [0.663, 0.722]. \end{aligned}$$

□

Exercise 9.9. Use the MATLAB function `integral_sum` on p. 132 from section 9.1 to compute S_1 through S_{10} in Example 9.7, using INTLAB's internal interval arithmetic. □

For the exact value of S_N in Example 9.7, we have the actual widths $w(S_N) = 1/N - 1/(2N) = 1/(2N)$, so the sequence $Y_{k+1}^* S_{k+1}^* \cap Y_k^*$ with $Y_1^* = S_1^*$ converges (using three-digit IA) in no more than 500 steps to an interval of width no less than 0.001 containing the exact value of the integral. This is slow convergence for such an easy integral; we will introduce faster methods for bounding integrals in this section. First, however, an example is needed to illustrate the application of (9.33) and (9.11).

Example 9.8. Suppose we wish to find upper and lower bounds on the integral

$$I = \int_0^1 e^{-t^2}\, dt. \tag{9.35}$$

For the integrand in (9.35) we have the interval polynomial enclosure P defined by

$$e^{-t^2} \in P(t) = 1 - t^2 + \tfrac{1}{2}t^4 - [0.0613, 0.1667]\, t^6 \tag{9.36}$$

for all $t \in [0, 1]$. The interval coefficient $[0.0613, 0.1667]$ bounds the range of values of $\frac{1}{6}e^{-s^2}$ for all $s \in [0, 1]$. From (9.33) we find, for I in (9.35),

$$I \in \int_{[0,1]} P(t)\, dt. \tag{9.37}$$

It follows from (9.12) and (9.36) that

$$I \in \int_{[0,1]} P(t)\,dt = \left[1 - \tfrac{1}{3} + \tfrac{1}{10} - \tfrac{.1667}{7},\ 1 - \tfrac{1}{3} + \tfrac{1}{10} - \tfrac{.0613}{7}\right] \subseteq [0.742, 0.758]. \quad (9.38)$$

By using narrower interval polynomial enclosures, we can, as we shall see, compute arbitrarily sharp bounds on the exact value of integrals such as (9.35). □

Using automatic differentiation, we can efficiently compute enclosures for integrals. From (9.21) and (9.6) we have the following theorem for any positive integers K and N, if a and b have the *same sign*.

Theorem 9.3.

$$\begin{aligned}\int_{[a,b]} f(t)\,dt \in \sum_{i=1}^{N} \Bigg\{ & \sum_{k=0}^{K-1} \frac{1}{k+1} (f)_k(\underline{X}_i) w(X_i)^{k+1} \\ & + \frac{1}{K+1} (f)_K(X_i) w(X_i)^{K+1} \Bigg\}\end{aligned} \quad (9.39)$$

for functions f with inclusion isotonic interval extensions of f and its first K derivatives. This includes the class $\mathrm{FC}_n(X_0)$.

Proof. From the Lipschitz property of the interval extensions, there is a constant L_K such that

$$w((f)_K(X_i)) \le L_K w(X_i) \text{ for all } X_i \subseteq [a,b] \subseteq X_0. \quad (9.40)$$

Let us denote the right-hand side of (9.39) by $I_{N,K}$. Since only the $(f)_K(X_i)$ contribute anything to the width of $I_{N,K}$, we have

$$\begin{aligned} w(I_{N,K}) &= w\left[\sum_{i=1}^{N} \frac{1}{K+1} (f)_K(X_i) w(X_i)^{K+1}\right] \\ &= \frac{1}{K+1} \sum_{i=1}^{N} w\left[(f)_K(X_i)\right] w(X_i)^{K+1} \\ &\le \frac{L_K}{K+1} \sum_{i=1}^{N} w(X_i) w(X_i)^{K+1} \qquad \text{by (9.40)} \\ &= \frac{L_K}{K+1} \sum_{i=1}^{N} w(X_i)^{K+2}. \end{aligned}$$

Hence,

$$w(I_{N,K}) \le \frac{L_K}{K+1} \sum_{i=1}^{N} w(X_i)^{K+2}. \quad (9.41)$$

For a uniform subdivision of $[a,b]$, we have $w(X_i) = (b-a)/N$ for $i = 1, \ldots, N$, and (9.41) becomes

$$w(I_{N,K}) \le C_K h^{K+1}, \quad \text{where} \quad C_K = \frac{L_K}{K+1}(b-a) \text{ and } h = \frac{b-a}{N}. \quad (9.42)$$

Thus, for fixed K, $I_{N,K}$ contains the exact value of the integral and has width no more than some constant C_K times the $(K+1)$st power of the "step size" h. □

Example 9.9. For (9.34), we can take $L_K = K+1$ (cf. Exercise 9.10), and hence $C_K = 1$ for all K. Thus, $I_{N,K}$ for the application of (9.39) to (9.34) is, for uniform subdivision, an interval of width no more than $(1/N)^{K+1}$ for any positive integers N and K. Furthermore, we have

$$(f)_k(\underline{X}_i) = (-1)^k/\underline{X}_i^{k+1}, \qquad (f)_K(X_i) = (-1)^K/X_i^{K+1},$$

with

$$\underline{X}_i = 1 + (i-1)/N, \qquad w(X_i) = 1/N, \qquad X_i = 1 + [i-1, i]/N.$$

To make the $w(I_{N,K}) \leq 0.001$, we can take $N \geq 10^{3/(K+1)}$. The simpler method (9.8) with width bounded by (9.10) corresponds to putting $K = 0$ in (9.39) and deleting the sum over k. The following pairs of integers satisfy $N \geq 10^{3/(K+1)}$.

K	N
0	1000
1	32
2	10
3	6
4	4
5	4
6	3
7	3
8	3
9	2

For $K > 9$ we still need $N = 2$.

Since the exact value of the integral is contained in $I_{N,K}$ for every pair of integers $N \geq 1$, $K \geq 0$, we can intersect the intervals $I_{N,K}$ for any sequence of pairs of integers $\{(N, K)\}$ and obtain an interval which still contains the exact value of the integral. If we carry out all the computations in $I_{N,K}$ using IA, then any sequence of such intersections will converge in some finite number of steps to an interval containing the exact value of the integral. We could fix N and take an increasing sequence of values of K, or fix K and take an increasing sequence of values of N, for instance. □

Exercise 9.10. Verify that we may take $L_K = K+1$ in Example 9.9. □

Example 9.10. For (9.35) we can also apply (9.39) to compute $I_{N,K}$ for any $N \geq 1$, $K \geq 0$, using the recursion formulas

$$\begin{aligned}(f)_0(X) &= e^{-X^2},\\ (f)_1(X) &= -2X(f)_0(X),\\ (f)_k(X) &= -(2/k)(X(f)_{k-1}(X) + (f)_{k-2}(X)) \text{ for } n \geq 2.\end{aligned} \tag{9.43}$$

For $N = 2$ and $K = 6$, using IA, we find $I_{N,K} = I_{2,6} = [0.746, 0.748]$. Compare with (9.38).

We can also apply (9.39) to more complicated functions in $FC_n(X_0)$, e.g. $x(t)$ in (9.30) using the recursion relations (9.32) for the derivatives.

Note that, during the computation of $I_{N,K}$ using (9.39) in IA, we must also compute the coefficients $(f)_k(\underline{X}_i)$ in IA to ensure that the computed interval value of $I_{N,K}$ contains the value defined by (9.39). □

9.5 Further Remarks on Interval Integration

A computer implementation of an interval integration routine typically finds the Taylor coefficients automatically, generates the interval enclosure of the associated remainder term, and adaptively chooses intervals of expansion within the interval of integration, until the entire integral is computed. If the data is given precisely, this can even be done to specified accuracy. If there is uncertainty in the data itself, for instance, if all we know about α is an interval that contains its value, we cannot expect to find a narrower interval containing the integral than is allowed by the uncertainty in the data. In any case, an interval enclosure of the integral still can be obtained.

Lang et al. [81, 129] as well as Krämer et al. [118] have successfully applied the techniques we have been describing to adaptive Gaussian quadrature, for rigorously verified enclosures of integrals.

We can also carry out interval integration over domains in more than one dimension, that is, we can compute *multiple integrals*. Let us briefly discuss interval integration over a simplex. A nondegenerate simplex S_d in $\mathbb{R}^d$ can be represented by

$$S_d = \left\{ x = \sum_{i=0}^{d} x_i p^{(i)} : 0 \le x_i \le 1, \ \sum_{i=0}^{d} x_i \le 1 \right\},$$

where $\{p^{(0)}, p^{(1)}, \ldots, p^{(d)}\}$ is the set of $d+1$ vertices. For $d = 3$, S_3 is a tetrahedron. The *barycentric coordinates* $\{x_i\}$ transform S_3 into the *standard* simplex with $p^{(0)} = 0$ and $p_j^{(i)} = \delta_{ij}$. So integration over S_d can be written as integration over the standard simplex. Integration over a tetrahedron can be written as

$$\int_{S_3} f(x, y, z)\, dx\, dy\, dz = \int_0^1 \int_0^{1-x_1} \int_0^{1-x_1-x_2} f(x_1, x_2, x_3)\, dx_3\, dx_2\, dx_1.$$

We can transform this problem further into integration over the unit cube as

$$\int_0^1 \int_0^1 \int_0^1 g(x_1, y_2, y_3)\, dy_3\, dy_2\, dx_1$$

by using the substitutions

$$x_3 = (1 - x_2 - x_1)y_3, \quad x_2 = (1 - x_1)y_2,$$

where

$$g(x_1, y_2, y_3) = (1 - x_2 - x_1)(1 - x_1) f(x_1, (1 - x_1)y_2, (1 - x_2 - x_1)y_3).$$

There is interval software for multiple integrals of more general types, and even allowing singular integrands (with certain restrictions), e.g., Aberth's RANGE software, when the integrand is precisely defined. A RANGE result, using interval methods, integration of Taylor series with interval remainder, plus adaptive subdivision of the region of integration, is *rigorous*. Although non-interval-based software packages will be able to compute many integrals (such as that in Exercise 9.11) accurately as well, they still lack the mathematical rigor of interval integration.

Exercise 9.11. Use RANGE to integrate $(x_1^2+x_2^2+x_3^2)e^{-x_1x_2x_3}$ over the standard tetrahedron S_3. (See the web page for this book, www.siam.org/books/ot110, to obtain RANGE.) □

We also mention some relations between the interval integral defined by (9.11) and other familiar types of integrals [33].

1. Let F_L and F_R be real-valued functions with common domain that includes an interval $[a, b]$, and such that $F_L(x) \leq F_R(x)$ for all $x \in [a, b]$. Let F be the interval-valued function defined by $F(x) = [F_L(x), F_R(x)]$. The interval integral $\int_a^b F(x)\,dx$ is equal to the interval from the *lower Darboux integral* of F_L to the *upper Darboux integral* of F_R:

$$\int_a^b F(x)\,dx = \left[\underline{\int_a^b} F_L(x)\,dx, \overline{\int_a^b} F_R(x)\,dx\right].$$

   The lower Darboux integral of a real-valued function $f(x)$ is the supremum of all the sums of step functions below $f(x)$ for any partition of the interval of integration, and the upper Darboux integral is the infimum of all the upper sums. The upper and lower Darboux integrals always exist in the extended real number system for *any* real-valued functions F_L and F_R. It follows that *all* interval-valued and hence *all* real- (degenerate interval-) valued functions are interval integrable.

2. The interval integral of a real- (degenerate interval-) valued function is a real number (degenerate interval) if and only if the function is Riemann integrable, which is equivalent to the upper and lower Darboux integrals being equal.

3. The interval integral of an interval-valued function F contains the Riemann integrals of all Riemann integrable functions in F, and the Lebesgue integrals of all Lebesgue integrable functions in F. Thus, $f(x) \in F(x)$ for all $x \in [a, b]$ implies that

$$\int_a^b f(x)\,dx \in \int_a^b F(x)\,dx,$$

   where the integral of f may be either the Riemann or Lebesgue integral as is appropriate for a given f. For example, the interval integral of the characteristic function of the rationals in the unit interval is the interval $[0, 1]$, which contains the Lebesgue integral, namely, zero.

9.6 Software and Further References

Use of automatic differentiation to compute Taylor polynomials in the way we have explained above was perhaps first suggested in [146], but it has been developed extensively since then under various guises. Often termed *Taylor arithmetic*, practicalities of the technique are discussed in [169]. Implementation of the univariate Taylor arithmetic we have described above is relatively simple, but, because of efficiency concerns, implementation of multivariate Taylor arithmetic (involving higher-order partial derivatives) is trickier. However, the COSY package of Berz et al. has a good implementation. In [23], Berz and Makino describe use of COSY's Taylor arithmetic for computing multiple integrals with verified bounds.

Although INTLAB implements automatic differentiation for multivariate functions, only first-order derivatives (with the `gradient` variable type) and second-order derivatives (with the `hessian` variable type) are available; no one has supplied a package for higher-order univariate Taylor arithmetic (using the formulas described above) yet. However, if the reader has access to Aberth's book [2], then the software supplied with that work contains a Taylor-arithmetic-based method for computing an integral with verified bounds.

More discussion of automatic differentiation in general can be found in [21, 38, 54, 200] and elsewhere. A recent reference for verified numerical integration is [191]. An older reference, [40] describes construction of a quadrature package along the lines we have outlined. Although this older package uses a system (ACRITH) that is no longer generally available, it still may be of interest. For C++ programmers, packages for automatic differentiation, verified quadrature, etc., are available with C-XSC; a description of the C-XSC system, as well as references for quadrature packages utilizing C-XSC, appears in [80].

In the next chapter, techniques from this chapter are combined with fixed-point iteration techniques (such as in Chapters 7 and 8) to introduce a few basic ideas for verified methods for solving integral equations. The next chapter also outlines a few basic techniques for the difficult problem of designing algorithms for rigorously bounding the solutions to ordinary and partial differential equations.

Chapter 10

Integral Equations and Differential Equations

10.1 Integral Equations

Recall that an interval enclosure of a real-valued function f is an inclusion isotonic interval-valued function F such that $f(t) \in F(t)$. Interval polynomial enclosures are particularly useful. Recall that we can formally integrate interval polynomials and that interval integration preserves inclusion. In this section, we will consider operator equations of the form

$$y(t) = p(y)(t), \tag{10.1}$$

where the operator p may include integrals of the function $y(t)$.

We will consider some interval methods for establishing the existence of solutions to (10.1) and computational tests for existence of solutions and for convergence of iterative methods for approximate solution of (10.1). For clarity of notation, we will restrict our attention in this section to equations of the form (10.1) in which we seek a real-valued function $y(t)$ of a single real variable t. The methods can be extended easily to the case of a vector-valued function of a real variable as in systems of ordinary differential equations, and some of the methods can be extended to cover vector-valued functions of vector-valued t (e.g., systems of partial differential equations).

If $X(t)$ and $Y(t)$ are interval- (or interval vector-) valued functions with a common domain, we write

$$X \subseteq Y \quad \text{if} \quad X(t) \subseteq Y(t) \text{ for all } t \text{ (in the common domain).} \tag{10.2}$$

Similarly, if $x(t)$ is a real- (or real vector-) valued function, we write

$$x \in X \quad \text{if} \quad x(t) \subset X(t) \text{ for all } t. \tag{10.3}$$

Suppose the operator p in (10.1) is defined for some class M_r of real-valued functions y with common domain $a \leq t \leq b$, and suppose $p\colon M_r \to M_r$. Let the *interval operator* $P\colon M \to M$ be defined on a class M of interval enclosures of elements of M_r with $M_r \subseteq M$. We call P an *interval majorant* of p if

$$p(y) \in P(Y) \text{ for } y \in Y. \tag{10.4}$$

An interval operator P is *inclusion isotonic* if

$$X \subseteq Y \quad \text{implies} \quad P(X) \subseteq P(Y). \tag{10.5}$$

We can usually write such an operator P immediately, given p. For example, if H and F are inclusion isotonic, then the interval operators of the form

$$P(Y)(t) = H(t, Y(t)) + \int_0^t F(t, s, Y(s))\, ds \tag{10.6}$$

are inclusion isotonic because of (9.13).

The following theorem provides a basis for useful computational tests for existence of solutions to (10.1) and for the convergence of iterative algorithms for solving operator equations.

Theorem 10.1. If P is an inclusion isotonic interval majorant of p, and if $P(Y^{(0)}) \subseteq Y^{(0)}$, then the sequence defined by

$$Y^{(k+1)} = P(Y^{(k)}) \qquad (k = 0, 1, 2, \dots) \tag{10.7}$$

has the following properties:

(1) $Y^{(k+1)} \subseteq Y^{(k)}$, $k = 0, 1, 2, \dots$.

(2) For every $a \le t \le b$, the limit

$$Y(t) = \bigcap_{k=0}^{\infty} Y^{(k)}(t) \tag{10.8}$$

exists as an interval $Y(t) \subseteq Y^{(k)}(t)$, $k = 0, 1, 2, \dots$.

(3) Any solution of (10.1) which is in $Y^{(0)}$ is also in $Y^{(k)}$ for all k and in Y as well. That is, if $y(t) \in Y^{(0)}(t)$ for all $a \le t \le b$, then $y(t) \in Y^{(k)}(t)$ for all k and all $a \le t \le b$;

(4) If there is a real number c such that $0 \le c < 1$, for which $X \subseteq Y^{(0)}$ implies

$$\sup_t w(P(X)(t)) \le c \sup_t w(X(t)), \qquad a \le t \le b, \tag{10.9}$$

for every $X \in M$, then (10.1) has the unique solution $Y(t)$ in $Y^{(0)}$ given by (10.8).

Proof. Property (1) follows by induction from $Y^{(1)} = P(Y^{(0)}) \subseteq Y^{(0)}$, using the inclusion isotonicity of P. For any fixed t, the sequence $Y^{(k)}(t)$ of nested intervals converges to an interval $Y(t)$ which is expressible as the intersection in property (2). If y is a solution of (10.1), then $y \in Y^{(0)}$ implies $p(y) \in P(Y^{(0)})$, since P is an interval majorant of p. However, $y = p(y)$, so $y \in Y^{(1)} = P(Y^{(0)})$. By induction, we obtain property (3). From (10.9), it follows that the limit $Y(t)$ in property (2) is a degenerate interval of zero width (real-valued) for every t, and $Y(t) = P(Y)(t)$. From (10.4), it follows that $Y(t)$ is a solution to (10.1). Uniqueness in $Y^{(0)}$ follows from the contraction property (10.9). □

10.2 ODEs and Initial Value Problems

A special case of the integral equation of the form

$$y(t) = h(t, y(t)) + \int_0^t f(t, s, y(s))\, ds$$

is the integral equation

$$y(t) = y_0 + \int_0^t f(s, y(s))\, ds,$$

which is formally equivalent to the *initial value problem* for the ODE

$$\frac{dy(t)}{dt} = f(t, y(t))$$

with initial condition

$$y(t) = y_0 \text{ at } t = 0.$$

A solution to the initial value problem is also a solution to the integral equation.

Example 10.1. The initial value problem

$$dy/dt = t^2 + y(t)^2, \qquad y(0) = 0, \tag{10.10}$$

can be written in the form (10.1) with

$$p(y)(t) = \int_0^t (s^2 + y(s)^2)\, ds. \tag{10.11}$$

We define in a natural way the interval operator P by

$$P(Y)(t) = \int_0^t (s^2 + Y(s)^2)\, ds. \tag{10.12}$$

Let $Y^{(0)}(t) = [0, w]$ for $0 \le t \le b$; then we have

$$P(Y^{(0)})(t) = \int_0^t (s^2 + [0, w^2])\, ds = t^3/3 + [0, w^2]t.$$

We have $P(Y^{(0)}) \subseteq Y^{(0)}$ if $b^3/3 + w^2 b \le w$. This is the case if, for instance, $w = 0.5$ and $b = 0.9$. Since P is (by construction) an inclusion isotonic majorant of p (see (5.24) and (9.13)), we can satisfy the hypotheses of Theorem 10.1 for this choice of w and b.

The operators p and P are defined for continuous functions y and Y. From (10.12), we find that, for $0 \le t \le b$ and $Y = [\underline{Y}, \overline{Y}] \subseteq Y^{(0)} = [0, w]$,

$$w(P(Y)(t)) = \int_0^t (\overline{Y}(s)^2 - \underline{Y}(s)^2)\, ds \le 2bw \sup_t w(Y(t)). \tag{10.13}$$

Therefore, (10.9) is satisfied with $c = 2bw$ if $2bw \le 1$. It turns out here that the b and w we found to satisfy $P(Y^{(0)}) \subseteq Y^{(0)}$, namely, $w = 0.5$ and $b = 0.9$, also satisfy (10.9). In other examples, we might have to reduce b to satisfy (10.9) after we have found w and

b to satisfy $P(Y^{(0)}) \subseteq Y^{(0)}$. It follows that the initial value problem (10.10) has a unique solution (expressible as the limit of the convergent sequence (10.7)) at least for $0 \le t \le 0.9$. We can apply the procedure again with $t = 0.9$ and $Y(0.9)$ as a new initial point to continue the solution beyond $t = 0.9$. □

Note that, even without (10.9), we will have convergence of the sequence (10.7) to some interval-valued function $Y(t)$.

Example 10.2. The initial value problem

$$dy/dt = \sqrt{y} \quad \text{for } y \ge 0, \text{ with the initial condition } y(0) = 0$$

can be written in the form (10.1) with

$$p(y)(t) = \int_0^t \sqrt{y(s)}\, ds.$$

We define the interval operator P by

$$P(Y)(t) = \int_0^t \sqrt{Y(s)}\, ds, \tag{10.14}$$

where

$$\sqrt{Y(s)} = \sqrt{\left[\underline{Y(s)},\ \overline{Y(s)}\right]} = \left[\sqrt{\underline{Y(s)}},\ \sqrt{\overline{Y(s)}}\right] \quad \text{for } 0 \le Y(s).$$

This operator does not satisfy (10.9) (cf. Exercise 10.2). Let $Y_0(t) = [0, w]$, with $w > 0$, for $0 \le t \le b$. Then

$$P(Y_0)(t) = \int_0^t \sqrt{[0, w]}\, ds = [0, \sqrt{w}]t.$$

We have $P(Y_0) \subseteq Y_0$ if $\sqrt{w}b \le w$. This is the case if $b \le \sqrt{w}$. As a numerical example, take $w = 1$ and $b = 1$. Then the sequence generated by (10.7) with $Y_0(t) = [0, 1]t$, $0 \le t \le 1$, is

$$Y^{(k+1)}(t) = P(Y^{(k)})(t) = \int_0^t \sqrt{Y^{(k)}(s)}\, ds \qquad (k = 0, 1, 2, 3, \ldots),$$

so

$$Y^{(1)}(t) = \int_0^t \sqrt{s}\, ds = \frac{2}{3} t^{3/2}[0, 1],$$

$$Y^{(2)}(t) = \int_0^t \sqrt{\frac{2}{3} s^{3/2}}\, ds = \frac{4}{21}\sqrt{t^{3/2}}\, t\sqrt{6} = \frac{4}{21} t^{7/4}\sqrt{6}[0, 1]$$

$$\vdots$$

By Theorem 10.1, the sequence is nested, has an interval-valued function

$$Y(t) = \textstyle\bigcap_{k=0}^{\infty} Y^{(k)}(t)$$

as a limit, and any solution $y(t)$ of $dy/dt = \sqrt{y}$, for $y \geq 0$, with the initial condition $y(0) = 0$ that is in $Y_0(t) = [0, 1]$ for $0 \leq t \leq 1$, is also in $Y^{(k)}(t)$ for every k and all $0 \leq t \leq 1$. Actually, the initial value problem has infinitely many solutions of the form

$$y_a(t) = \begin{cases} 0, & 0 \leq t \leq a, \\ \frac{1}{4}(t-a)^2, & a < t, \end{cases}$$

where a is an arbitrary nonnegative real number, and we have $y_a(t) \in Y^{(k)}(t)$ for all k. □

Exercise 10.1. Show that

$$Y(t) = \bigcap_{k=0}^{\infty} Y^{(k)}(t) = \left[0, \tfrac{1}{4} \sup_{0 \leq a \leq t} \{(t-a)^2\}\right] = \left[0, \tfrac{1}{4}t^2\right]. \qquad \square$$

Exercise 10.2. Show that the interval operator P given by (10.14) does not satisfy (10.9). □

In the previous example, $f(y) = \sqrt{y}$ was not differentiable at the initial condition $y(0) = 0$ because $f'(y) = 1/(2\sqrt{y})$ is not defined as a real number for $y = 0$, so the solution was not unique. Nevertheless, we were able to use interval methods to enclose the set of all infinitely many solutions. Let us now consider more typical examples, when the functions in a differential equation or system of differential equations are differentiable as many times as wanted, so we can use Taylor expansions with interval bounding of remainder terms and automatic differentiation to generate coefficients, as discussed in section 9.3.

Example 10.3. The initial value problem

$$dy/dt = t^2 + y^2, \quad y(0) = 1, \tag{10.15}$$

can be written as the integral equation

$$y(t) = 1 + \int_0^t (s^2 + y(s)^2)\, ds,$$

and we can define the interval operator P by

$$P(Y)(t) = 1 + \int_0^t (s^2 + Y(s)^2)\, ds.$$

We can choose an initial constant interval value, say, $Y_0(t) = [0, 2]$, and seek a positive number b such that $P(Y_0)(t) \subseteq Y_0(t)$ for all $t \in [0, b]$, that is,

$$1 + \frac{t^3}{3} + [0, 4]\, t \subseteq [0, 2] \text{ for all } t \in [0, b].$$

This is true if $1 + b^3/3 + 4b \leq 2$. We can determine in various ways (including by numerically solving the polynomial equation $b^3/3 + 4b - 1 = 0$) that this is true, for example, if $b = 0.24$, since

$$1 + 0.24^3/3 + 4(0.24) = 1.964608 \leq 2.$$

It follows that the initial value problem has a solution such that $y(t) \in [0, 2]$ for all $t \in [0, 0.24]$. It is not hard to show that P in this example satisfies (10.9), so the solution is unique, and the sequence (10.7) converges to it.

Alternatively, since the formal integrations get a bit messy, we can compute an enclosure of the solution $y(t)$ of the form (9.21) with interval enclosure of the remainder. To illustrate, we choose $N = 5$. Then

$$y(t) \in \sum_{k=0}^{4} (y)_k t^k + R_5([0, 0.24])t^5 \text{ for all } t \in [0, 0.24].$$

The recursive formulas given in Chapter 9, for this example, become

$$\begin{aligned}
(y)_1 &= dy/dt = t^2 + y^2, \\
(y)_2 &= \frac{1}{2}(2t + 2(y)_0(y)_1), \\
(y)_3 &= \frac{1}{3}(1 + 2(y)_0(y)_2 + (y)_1^2), \\
(y)_4 &= \frac{1}{4}(2(y)_0(y)_3 + 2(y)_1(y)_2), \\
(y)_5 &= \frac{1}{5}(2(y)_0(y)_4 + 2(y)_1(y)_3 + (y)_2^2).
\end{aligned}$$

From the initial condition at $t = 0$, we have $(y)_0 = y(0) = 1$. We find the initial coefficients

$$(y)_1 = 1, \quad (y)_2 = 1, \quad (y)_3 = \frac{4}{3}, \quad (y)_4 = \frac{7}{6}.$$

It remains to find $R_5([0, 0.24])$ such that $(y)_5(s) \in R_5([0, 0.24])$ for all $s \in [0, 0.24]$. For this, we recall that $Y_0(t) = [0, 2]$, so for $s \in [0, 0.24]$ and $y(s) \in [0, 2]$, using interval arithmetic, we find

$$\begin{aligned}
(y)_0 &= y(s) \in [0, 2], \\
(y)_1 &= dy/dt = s^2 + y^2 \\
&\in [0, 0.24]^2 + [0, 2]^2 \subset [0, 4.1], \\
(y)_2 &= s + y(y)_1 \\
&\in [0, 0.24] + [0, 2][0, 4.1] \subset [0, 8.44], \\
(y)_3 &= \frac{1}{3}(1 + 2y(y)_2 + (y)_1^2) \\
&\in \frac{1}{3}(1 + 2[0, 2][0, 8.44] + [0, 4.1]^2) \subset [0.33, 17.2], \\
(y)_4 &= \frac{1}{4}(2(y)_0(y)_3 + 2(y)_1(y)_2) \\
&\in \frac{1}{2}([0, 2][0.33, 17.2] + [0, 4.1][0, 8.44]) \subset [0, 34.51], \\
(y)_5 &= \frac{1}{5}(2(y)_0(y)_4 + 2(y)_1(y)_3 + (y)_2^2),
\end{aligned}$$

so

$$(y)_5(s) \in R_5([0, 0.24])$$
$$\subset \frac{1}{5}(2[0, 2][0, 34.51] + 2[0, 4.1][0.33, 17.2] + [0, 8.44]^2) \text{ for all } s \in [0, 0.24].$$

Thus, we can take $R_5([0, 0.24]) = [0, 70.1]$, and we have the result that the initial value problem (10.15) has a unique solution $y(t)$ at least for $t \in [0, 0.24]$ with an interval polynomial enclosure $Y(t)$ given by

$$y(t) \in Y(t) = 1 + t + t^2 + \frac{4}{3}t^3 + \frac{7}{6}t^4 + [0, 70.1]\,t^5 \text{ for all } t \in [0, 0.24].$$

In particular,

$$y(0.1) \in [1.1114, 1.1122] = 1.1118 \pm 4(10^{-4}).$$

Aberth's RANGE finds $y(0.1) = 1.11146\tilde{}$, which means that the exact solution is, in fact, contained in the interval we found.[39]

We show the calculations in detail above to give an idea of how some of the software works. What is shown is a relatively simple version of much more sophisticated but similar techniques in existing software. It is important to point out that everything shown can be carried out automatically by software, given only the differential equations and initial conditions, and at whatever values of t the solution is desired.

We can continue an interval enclosure, such as the $Y(t)$ found above, for further values of the independent variable $t > 0.24$ by restarting the procedure with a new initial condition, such as $y(0.24) \in [1.1114, 1.1122]$. □

The interval Taylor method extends to systems of differential equations and allows for intervals of initial conditions and parameters in the functions.

Exercise 10.3. Apply the interval Taylor method to the following initial value problems: (a) $dy/dt = y^2$ with $y(0) = 1$ and (b) $dy/dt = -y$ with $y(0) = 1$. □

A difficulty arises in connection with the continuation of sets of solutions to an initial value problem for a system of n ordinary differential equations. Suppose a set of initial points $x(0)$ is enclosed in a finitely representable set S_0 such as an interval vector, an ellipsoid, or a polytope. The set of solution points

$$S_t = \{x(t) = (x_1(t), \ldots, x_n(t)) \colon x(0) \in S_0\}$$

for $t > 0$ is, in general, not exactly representable by the same type of geometrical object as S_0. Thus if S_t is enclosed again in an interval vector, ellipsoid, or polytope, the resulting bounded region will contain extra points, not in the solutions emanating from S_0. This phenomenon has been dubbed *the wrapping effect* [152] and has been studied in a variety of ways. Successful attempts to reduce the growth of bounds due to the wrapping effect have been based on various ideas, the most successful of which is the Taylor model approach of Berz and Makino. See, for instance, [22, 24, 25, 138, 139]. Earlier references on interval methods for ODEs can be found in [146, 147, 148, 152] as well as by searching the web.

[39] Here, the tilde means that the exact result is 1.11146 ± 0.000005.

10.3 ODEs and Boundary Value Problems

A two-point boundary value problem of the form

$$\frac{d^2y}{dt^2} = f(t, y(t)), \qquad y(0) = y(1) = 0,$$

is equivalent to the integral equation

$$y(t) = p(y)(t) = (t-1)\int_0^t sf(s, y(s))\,ds + t\int_t^1 (s-1)f(s, y(s))\,ds.$$

Exercise 10.4. Verify this. □

If F is an interval enclosure of f, the interval operator P defined by

$$P(Y)(t) = (t-1)\int_0^t sF(s, Y(s))\,ds + t\int_t^1 (s-1)F(s, Y(s))\,ds$$

is an interval majorant of p, and we can apply Theorem 10.1.

Many published papers on this topic can be found with a web search. Examples are [107, 133, 187].

10.4 Partial Differential Equations

In the area of *partial differential equations*, IA has been used in computer-aided rigorous computation of error bounds, as well as for obtaining mathematically rigorous results in eigenvalue problems and the analysis of operators. See, e.g., [113, 164, 166, 193, 231, 242, 250].

Details are well beyond the scope of this introductory volume. Much remains to be done in the area of interval methods for partial differential equations.

For additional examples of the application of interval analysis to operator equations, as well as a brief discussion of interval analysis in the context of lattice theory, see Moore and Cloud [156].

In our final chapter, we discuss applications of the concepts and techniques introduced to this point. Some of these applications, such as computer-assisted proofs and global optimization, are general, with many specific examples. Our coverage of these applications is necessarily uneven, based on our own expertise. Also, detailed treatment of the derivation of these applications is outside the scope of this book. The interested reader should pursue the references we provide to obtain a fuller understanding of particular applications.

Chapter 11

Applications

11.1 Computer-Assisted Proofs

Any computation using IA, and rigorous methods of interval analysis, proves something. The intervals found by the computer will, by construction, certainly contain the results they are constructed to contain.

Because of the rigor of interval computation, it is finding use in computer-aided proofs in mathematical analysis, among other areas [42]. Interval analysis has been used, for example, in computational parts of a proof of Kepler's Conjecture on the densest packing of spheres. The following is quoted from Szpiro [240]:

> The fascinating story of a problem that perplexed mathematicians for nearly 400 years. In 1611, Johannes Kepler proposed that the best way to pack spheres as densely as possible was to pile them up in the same way that grocers stack oranges or tomatoes. This proposition, known as Kepler's Conjecture, seemed obvious to everyone except mathematicians, who seldom take anyone's word for anything. In the tradition of Fermat's Enigma, George Szpiro shows how the problem engaged and stymied many men of genius over the centuries—Sir Walter Raleigh, astronomer Tycho Brahe, Sir Isaac Newton, mathematicians C. F. Gauss and David Hilbert, and R. Buckminster Fuller, to name a few—until Thomas Hales of the University of Michigan submitted what seems to be a definitive proof in 1998.

Another proof that used interval arithmetic for rigor was that of the "double bubble conjecture." It had been conjectured that two equal partial spheres sharing a boundary of a flat disk separate two volumes of air using a total surface area that is less than any other boundary. This equal-volume case was proved by Hass et al. [71], who reduced the problem to a set of 200,260 integrals, which they carried out on an ordinary PC.

Warwick Tucker has given a rigorous proof that the Lorenz attractor exists for the parameter values provided by Lorenz. This was a long-standing challenge to the dynamical system community and was included by Smale in his list of problems for the new millennium. The proof uses computer estimates with rigorous bounds based on interval analysis.

In later work, Warwick Tucker made further significant contributions to the development and application of this area [244].

In a chapter on computer-assisted proofs in Einarsson's book [45], Siegfried Rump listed the following examples of computer-assisted proofs, all of which make use of interval methods:

- verification of the existence of the Lorenz attractor,
- verification of the existence of chaos,
- double-bubble conjecture,
- verification of the instability for the Orr–Sommerfeld equations with a Blasius profile,
- dynamics of the Jouanolou foliation,
- solution of the heat convection problem,
- verified bounds for the Feigenbaum constant,
- existence of an eigenvalue below the essential spectrum of the Sturm–Liouville problem,
- eigenfrequencies of a turbine,
- SPICE program for circuit analysis,
- extreme currents in Lake Constance,
- forest planning,
- global optimization,
- all the zeros of a nonlinear system in a box,
- least squares problems.

See also Meyer and Schmidt [143], in which more than half the papers involve the use of interval methods, including

- computer-assisted approach to small divisors problems arising in Hamiltonian dynamics,
- computer-assisted proofs of stability of matter,
- accurate strategies for Kolmogorov–Arnold–Moser (K.A.M.) bounds and their implementation,
- software tool for analysis in function spaces,
- interval tools for computer-aided proofs in analysis,
- computer-assisted lower bounds for atomic energies.

11.2 Global Optimization and Constraint Satisfaction

It has long been asserted that rigorous global optimization by a computer is impossible in the nonlinear case, and this may be true using only evaluations of functions at points. At the very least we need to be able to compute upper and lower bounds on ranges of values of functions over sets. IA, as we have seen, provides upper and lower bounds on ranges of values of functions over continua, including intervals and vectors of intervals. This fact alone provides a basic algorithm for global optimization without using derivatives, which we describe in this section. If the objective function and the constraint functions are differentiable, there are more efficient interval methods using, for instance, the interval, Newton method, but we will not cover those here. See the references at the end of this section for more efficient methods.

For an n-dimensional region ("box") such as an interval vector

$$B = (B_1, B_2, \ldots, B_n)$$

using IA, we can test, on a computer, the truth of a relation such as

$$f(x) = f(x_1, x_2, \ldots, x_n) \leq 0$$

for *all* points $x = (x_1, x_2, \ldots, x_n) \in B$, where f is a real-valued function with an inclusion isotonic interval extension F. For example, if $F(B) = [L, U] \leq 0$, that is, if $U \leq 0$, then $f(x) \leq 0$ for all $x \in B$.

We consider now the following *global optimization problem*:

$$\left\{ \begin{array}{l} \text{minimize } f(x) \\ \quad \text{subject to the } \textit{constraints} \\ \qquad\quad x \in B \text{ and} \\ \qquad\quad p_i(x) \leq 0, \ i = 1, 2, \ldots, k. \end{array} \right. \tag{11.1}$$

Points satisfying the constraints are called *feasible*.

That is, we wish to find the set X^* of all global minimizers $x^* \in B$ and the global minimum value f^* such that

$$f(x^*) = f^* \leq f(x)$$

for all feasible points x.

11.2.1 A Prototypical Algorithm

The following simple algorithm will find a list of small boxes whose union contains the set X^* of all feasible global minimizers, along with an interval containing f^*. It is a simple branch-and-bound algorithm, using interval computation for rigor and completeness.

The algorithm proceeds by deleting parts of the initial box B which cannot contain a global minimizer, leaving a list of subboxes whose union still contains the set X^* of all global minimizers. During this process, we obtain upper and lower bounds on f^*.

In the algorithm, X is any subbox of B, which may even be a point (degenerate box) in B. We assume we have inclusion isotonic interval extensions, F of f and P_i of p_i, defined on B, and hence defined on subboxes of B. There are two kinds of *deletion* tests on X made in the algorithm:

1. Feasibility test.

   (a) If $P_i(X) \leq 0$ for all $i = 1, 2, \ldots, k$, then X *is certainly feasible*.
   That every point in X is feasible is guaranteed despite rounding errors in the computer floating point arithmetic because of the outward rounding used in IA. Thus $P_i(X) \leq 0$ implies that $p_i(x) \leq 0$ for all $x \in X$.

   (b) If for some i, $P_i(X) > 0$, then X *is certainly infeasible* (contains no feasible points), *delete* X.

2. Midpoint test.

   If $m(X)$ is feasible, we compute $F(m(X))$. Denote this interval by $[LFMX, UFMX] = F(m(X))$. It is certainly the case that $f^* \leq UFMX$, because $f^* \leq f(m(X)) \leq UFMX$. If Y is another subbox of B, and we evaluate $F(Y)$ to find $[LFY, UFY] = F(Y)$, then we can make the following test. Let UF^* be the smallest $UFMX$ yet found.

   (a) If $LFY > UF^*$, then Y cannot contain a feasible global minimizer in B, and we *delete* Y.

Using tests 1 and 2, we can construct a list (queue) of *items* consisting of pairs (X, LFX), where $F(X) = [LFX, UFX]$. We LIST items at the end and UNLIST items from the beginning of the list.

Every time we UNLIST a box, we bisect it in a coordinate direction of maximum width. We test each half. If it is not deleted, we put it at the end of the list. As a result, the widest box remaining is always the first one in the list. If it is narrower than some prespecified tolerance `EPSB` (epsilon B), then so are all the other remaining subboxes, and we terminate the algorithm. The bisections are always in a coordinate direction of maximum width.

Algorithm 11.1.

initialize the list with the single item (B, LFB) and set $UF^* = UFB$

DO WHILE (the first box on the list has width $>$ `EPSB`.)

unlist first box on the list

bisect $X = X_1 \cup X_2$

delete X_1 if 1(b) applies for $X = X_1$ or if 2(a) applies for $Y = X_1$

otherwise list (X_1, LFX_1) at end of list and update $UF^* = UFX_1$, if MX_1 is feasible and UFX_1 reduces UF^*

delete X_2 if 1(b) applies for $X = X_2$ or if 2(a) applies for $Y = X_2$

otherwise list (X_2, LFX_2) at end of list and update $UF^* = UFX_2$ if MX_2 is feasible and UFX_2 reduces UF^*

END DO

output box list items

$LF^* = \min(LFX)$, over the finite set of all items (X, LFX) in the list

UF^* is the current UF^*

stop

When we stop, we will have $LF^* \leq f^* \leq UF^*$, and the union of the boxes X in the list of items (X, LFX) will certainly contain all the feasible global optimizers of f in B.

This algorithm has appeared in Moore et al. [158]. There, various improvements in efficiency are discussed for differentiable problems, including use of the interval Newton method. Also, it is shown how to parallelize the procedure for further improvement in efficiency with application to photoelectron spectroscopy.

Algorithm 11.1, while not efficient for differentiable problems, is applicable in general. In [158], the following nondifferentiable example is discussed:

minimize

$$f(x) = (|x_1^2 + x_2^2 - 1| + 0.001)|x_1^2 + x_2^2 - 0.25|$$

subject to

$$p(x) = \max\{(1 - \max\{\tfrac{|x_1|}{0.6}, \tfrac{|x_2|}{0.25}\}), (1 - \max\{\tfrac{|x_1|}{0.25}, \tfrac{|x_2 - 0.4|}{0.3}\})\} \leq 0$$

and $x = (x_1, x_2) \in B = ([-1.2, 1.2], [-1.2, 1.2])$.

The objective function f and constraint function p are nondifferentiable and nonconvex, and the feasible region is nonconvex.

The set X^* of all global minimizers consists of three disconnected continua, three disjoint arcs on the circle $x_1^2 + x_2^2 = 0.25$. The arcs lie outside the union of two rectangles cutting through the circle. The entire unit circle $x_1^2 + x_2^2 = 1$ consists of local minimizers. The local, but not the global, minimizers are eliminated by choosing the final box width tolerance `EPSB` ≤ 0.1. The algorithm produces a list of small boxes whose union covers the set X^* of all global minimizers. For `EPSB` $= 0.0125$, the paper cited shows the graph of X^*. It is the familiar "smiley face."

11.2.2 Parameter Estimation

Simple variations of Algorithm 1 can solve other types of problems. For example consider the following variation. Here, 1(a) and 1(b) refer to the feasibility test on p. 160. The bisections are, as before, in a coordinate direction of maximum width.

Algorithm 11.2.

initialize the list with the single item B, and max box width for termination `EPSB`.

DO WHILE (the first box on the list has width $>$ `EPSB`.)

unlist first box on the list

bisect $X = X_1 \cup X_2$

delete X_1 if 1(a) or 1(b) applies for $X = X_1$

otherwise list X_1 at end of list

delete X_2 if 1(a) or 1(b) applies for $X = X_2$

otherwise list X_2 at end of list

END DO

output box list items

stop

The union of the sets represented by the boxes output at the termination of the algorithm will contain all the *boundary points* of the set of all feasible points $\{x\colon P_i(x) \le 0 \text{ for all } i = 1, 2, \ldots, k\}$, because the only parts of B that are deleted by Algorithm 11.2 are subboxes containing only all feasible or all infeasible points. Any point on the boundary of the feasible region will contain both feasible and infeasible points in any neighborhood. Such points will all lie in the union of the remaining subboxes.

Algorithm 11.2 has been applied to estimating the set of all parameters of a model for which the output errors are within given bounds. See Moore [153]. A more thorough discussion of related methods, with applications to robust control and robotics, can be found in Jaulin et al. [86].

11.2.3 Robotics Applications

Although there are various global optimization problems and problems involving the solution of nonlinear systems that we do not yet know how to solve efficiently with interval techniques, we have been largely successful with applications involving robotics. An early such application was an example from [243] and has since become a classic test problem for interval methods for nonlinear systems. The mathematical formulation consists of finding all solutions to $f(x) = 0$, where $f\colon \mathbb{R}^8 \to \mathbb{R}^8$, and the components of f are defined by

$$
\begin{aligned}
f_1 &= \alpha_1 x_1 x_3 + \alpha_2 x_2 x_3 + \alpha_3 x_1 + \alpha_4 x_2 + \alpha_5 x_4 + \alpha_6 x_7 + \alpha_7,\\
f_2 &= \alpha_8 x_1 x_3 + \alpha_9 x_2 x_3 + \alpha_{10} x_1 + \alpha_{11} x_2 + \alpha_{12} x_4 + \alpha_{13},\\
f_3 &= \alpha_{14} x_6 x_8 + \alpha_{15} x_1 + \alpha_{16} x_2,\\
f_4 &= \alpha_{17} x_1 + \alpha_{18} x_2 + \alpha_{19},\\
f_5 &= x_1^2 + x_2^2 - 1,\\
f_6 &= x_3^2 + x_4^2 - 1,\\
f_7 &= x_5^2 + x_6^2 - 1,\\
f_8 &= x_7^2 + x_8^2 - 1,
\end{aligned}
\tag{11.2}
$$

where

$$\begin{array}{lll}
\alpha_1 = 4.731 \cdot 10^{-3}, & \alpha_2 = -3.578 \cdot 10^{-1}, & \alpha_3 = -1.238 \cdot 10^{-1}, \\
\alpha_4 = -1.637 \cdot 10^{-3}, & \alpha_5 = -9.338 \cdot 10^{-1}, & \alpha_6 = 1.000, \\
\alpha_7 = -3.571 \cdot 10^{-1}, & \alpha_8 = 2.238 \cdot 10^{-1}, & \alpha_9 = 7.623 \cdot 10^{-1}, \\
\alpha_{10} = 2.638 \cdot 10^{-1}, & \alpha_{11} = -7.745 \cdot 10^{-2}, & \alpha_{12} = -6.734 \cdot 10^{-1}, \\
\alpha_{13} = -6.022 \cdot 10^{-1}, & \alpha_{14} = 1.000, & \alpha_{15} = 3.578 \cdot 10^{-1}, \\
\alpha_{16} = 4.731 \cdot 10^{-3}, & \alpha_{17} = -7.623 \cdot 10^{-1}, & \alpha_{18} = 2.238 \cdot 10^{-1}, \\
\alpha_{19} = 3.461 \cdot 10^{-1}, & &
\end{array}$$

and all solutions are known to lie within the unit 8-cube $x_i \in [-1, 1]$, $1 \le i \le 8$. (The four pairs $\{(x_{2i-1}, x_{2i})\}_{i=1}^{4}$ represent the cosines and sines of angles on the arm of the manipulator.)

Problem (11.2) is interesting from the standpoint that, depending on the actual branch-and-bound code used, several solutions can be missed if machine interval arithmetic without directed rounding is used. This problem was solved effectively with the early Fortran 77 interval code that later became INTBIS [105] with an efficiency that competed well with homotopy methods for polynomial systems; see [94]. The authors of such homotopy methods subsequently improved the selection of homotopy, scaling, and path-following algorithms, partially in response.

Our GlobSol code [104] finds rigorous enclosures to all 16 solutions to problem (11.2) with lower and upper bounds differing beyond the fifth digit (and a mathematical proof that there are no other solutions) with a total of only 31 boxes processed in the branch-and-bound process and 17 calls to a local approximate root-finder.

More recent applications to robotics can be seen in Chapter 8 of [86] and in [131]. In [131], "the geometric design problem of serial-link robot manipulators with three revolute (R) joints is solved for the first time using an interval analysis method." Their implementation of an interval branch-and-bound algorithm involved the ALIAS library, designed for parallel computation [142]. The problem was solved using a network of 25 desktop computers running in parallel.

The papers [192] and [34] are two other examples of applications to control and robotics.

11.2.4 Chemical Engineering Applications

Chemical engineering is another area where there have been major successes in the application of interval computations, largely because of the talents of Mark Stadtherr and his students. Much of this success is along the lines of optimizing objectives and solving systems of equations arising in molecular models, chemical kinetics equilibrium problems, and the like; see [48, 49, 236, 237]. Apparently, such problems often can be posed in a way that is amenable to solution by interval methods. Particularly impressive is the work of Lin and Stadtherr [134, 135, 136, 137], in which the authors combine Taylor arithmetic for the solution of systems of ordinary differential equations with interval methods for global optimization to rigorously determine parameters in differential equation models of processes.

A classic kinetics equilibrium problem, originally from [141], now not so challenging for interval techniques, consists of finding all solutions to $f(x) = 0$, where $f\colon \mathbb{R}^4 \to \mathbb{R}^4$ and the components of f are defined by

$$\begin{aligned} f_1 &= \alpha_1 x_2 x_4 + \alpha_2 x_2 + \alpha_3 x_1 x_4 + \alpha_4 x_1 + \alpha_5 x_4 \\ f_2 &= \beta_1 x_2 x_4 + \beta_2 x_1 x_3 + \beta_3 x_1 x_4 + \beta_4 x_3 x_4 + \beta_5 x_3 + \beta_6 x_4 + \beta_7 \\ f_3 &= x_1^2 - x_2 \\ f_4 &= x_4^2 - x_3 \end{aligned} \tag{11.3}$$

with

$$\begin{aligned} \alpha_1 &= -1.697 \cdot 10^7, & \alpha_2 &= 2.177 \cdot 10^7, & \alpha_3 &= 0.550 \cdot 10^0, \\ \alpha_4 &= 0.450 \cdot 10^0, & \alpha_5 &= -1.000 \cdot 10^0, \end{aligned}$$

and

$$\begin{aligned} \beta_1 &= 1.585 \cdot 10^{14}, & \beta_2 &= 4.126 \cdot 10^7, & \beta_3 &= -8.285 \cdot 10^6, \\ \beta_4 &= 2.284 \cdot 10^7, & \beta_5 &= -1.918 \cdot 10^7, & \beta_6 &= 4.840 \cdot 10^1, \\ \beta_7 &= -2.773 \cdot 10^1. \end{aligned}$$

The search box is set to $x_i \in [-0.01, 1.01]$, $1 \le i \le 4$. GlobSol finds a box with coordinate endpoints differing in the fourth digit, processing a total of nine boxes in the branch-and-bound process, and with three calls to a local approximate root finder.

11.2.5 Water Distribution Network Design

Angelika Hailer has recently presented a dissertation [55] in which she both provides a more accurate model of water distribution networks than previously found in the literature and successfully designs a global optimization algorithm and software package that provide automatically verified optima of this model.

11.2.6 Pitfalls and Clarifications

It is well known that the general global optimization problem (11.1) is NP-complete, that is, that there is no known algorithm that will solve all possible instances of this problem, even if the objective function f is quadratic and the p_i are linear, in a time that is bounded by a polynomial function of the number of variables and constraints. Because of these theoretical results, we cannot expect any software designed to solve this problem in general to be practical for all such problems. This holds for all solvers that claim to find an approximation to the global optimum within a specified error tolerance, regardless of whether interval arithmetic is used to bound roundoff error and provide mathematically rigorous results.

In fact, it is not hard to find global optimization problems that cannot be solved efficiently with existing software claiming to find solutions with automatic result verification. An example of this is found on Siegfried Rump's website, listed on the web page for this book.[40]

Nonetheless, there are notable additional successes (not mentioned above) in finding verified solutions to global optimization problems, based on carefully defining and limiting what we mean by "verified" or taking full advantage of the special structure of the problems. Before we enumerate these additional successes, we will clarify what a "verified solution" can mean. Doing so, we distinguish between what operations researchers call a *feasible solution*, an *optimum value* of the objective function f, and an *optimizer*:

A global optimum is a value of $\underline{f}$ of f such that $\underline{f} \le f(x)$ for every feasible point x.

A feasible solution is any point x for which $x \in B$ and $p_i(x) \le 0$, for every i, but is not necessarily a point at which f takes on its global minimum value. Feasible solutions are of interest in operations research problems for which an operating scheme that is possible to carry through is needed, but for which it may be difficult or impractical to find such a point at which f takes on its global optimum.[41]

A global optimizing point is a feasible point x such that $f(x)$ is the global optimum.

With this terminology, we can clarify different interpretations of the statement "solve the global optimization problem," that is, we define varying specifications for a computer program that deals with (11.1):

1. Find all feasible points x at which $f(x) = \underline{f}$, and output a list $\mathcal{C}$ of narrow boxes X such that the computer has proven that every feasible point x at which $f(x) = \underline{f}$ is in some $X \in \mathcal{C}$.

2. Find the global optimum $\underline{f}$ to within some specified tolerance ϵ, and find a box X within which we can prove there is a feasible point x such that $f(x) - \underline{f} < \epsilon$.

3. Output a small box X that is proved to contain a feasible point whose objective value is within some tolerance ϵ of the global optimum.

4. Output a small box X that has been proved to contain a feasible point x.

Specification 1 is the most difficult to meet and is likely to be practical to meet for the fewest number of problems. In computer programs whose goal is to meet specification 1, implementations are higher quality if the total number of boxes in $\mathcal{C}$ is minimal, there are no $X \in \mathcal{C}$ that do not contain any global optimizing points, each isolated global optimizing point is in a separate $X \in \mathcal{C}$, and the width of each component of each $X \in \mathcal{C}$ is as small as possible. Except for a very restricted set of problems, these quality conditions are impossible to meet in practice,[42] so we must settle for some boxes X that might not contain feasible points or optimizing points and boxes X that are not optimally narrow. In a computer

[40] www.siam.org/books/ot110.

[41] A term common to mathematical literature on optimization, a "feasible solution" is not a solution to the optimization problem but is simply a feasible point, that is, a point at which the constraints are satisfied.

[42] Using fixed-precision machine arithmetic, or even using variable precision and a finite amount of time.

code whose goal is to meet specification 1, if the code exits with a list $\mathcal{C}$, then we may graph two-dimensional cross sections of $\mathcal{C}$; these graphs typically exhibit clusters of boxes about isolated solution points and solution sets, and the widths of these clusters decrease by decreasing the stopping tolerances.[43] The "clustering effect" and ways of reducing the total number of boxes in such clusters are discussed first in [44] and [102], then in [228].

Software based on specification 2 is practical for a significantly wider set of problems than software based on specification 1. In particular, for certain nonsmooth problems and ill-conditioned or ill-posed problems, the algorithm may quit when $UF^* - LF^* < \epsilon$ and the box containing the point x at which $f(x) = UF^*$ may be printed; the other boxes need not be further resolved or verified. Christian Keil and Christian Jansson [108] have been successful at handling a variety of difficult test problems from the Netlib LP test set[44] with this criterion. Basically, this process first uses an efficient commercial or open-source solver to compute an approximate solution with floating point arithmetic. The approximate primal and dual variables then enter a short interval computation[45] to obtain upper and lower bounds on the optimum value.[46] The approximate optimizing point is then perturbed into the interior of the feasible region similarly to the techniques in [98], a small box is constructed about the perturbed point, and interval evaluation of the inequality constraints combined with the Krawczyk or other interval Newton method applied in a subspace for the equality constraints, proves existence of a feasible point within that small box. Interval evaluation of the objective over the tiny box obtained from applying the Krawczyk or interval Newton method then supplies an upper bound that is close to the optimum objective value. See [108] for the numerical results. Also see [85] for application of similar techniques in semidefinite programming.

Specification 2 can be met in the cases just outlined because a good approximation to the optimum and at least one optimizing point can usually be found by a noninterval solver, and theory particular to these cases allows us to use these approximations to obtain rigorous bounds. However, linear programs are often ill-posed in the sense that there is an entire manifold of optimizing points. In such instances, specification 2 and specification 1 lead to different results, even for linear problems, and trying to meet specification 1 is much more difficult than trying to meet specification 2. A discussion, as well as some techniques relevant to this case, appears in [100].

Specification 3 is essentially practical for most instances of the optimization problem (11.1) for which the form of the objective f and constraints p_i are sufficiently simple or can be put into a form in which there is not excessive overestimation in the interval evaluations over boxes within a few orders of magnitude of the machine precision, for which a sufficiently good approximation to a local optimizer can be found[47] and for which the local optimizer is isolated and for which the Kuhn–Tucker matrix[48] is sufficiently well-conditioned at the local optimizer. Essentially, to meet this specification, all that needs to be done is to construct an appropriately sized box about the local optimizer and perform the Krawczyk method or some other interval Newton method.

[43] With corresponding increases in computational time.
[44] See the web page for our book, www.siam.org/books/ot110.
[45] Involving only several interval matrix-vector multiplications.
[46] Neumaier and Shcherbina [173] and Jansson [84] discovered this process independently.
[47] Using, for example, a floating point nonlinear programming solver such as IPOPT from the COIN project.
[48] That is, the matrix of partial derivatives of the equations formed from the Lagrange multiplier conditions.

Specification 4 is the easiest to meet: we simply need to find an approximate feasible point (that is, a "feasible solution"), then use techniques such as those in [98] to construct bounds within which we prove that an exact feasible point exists and for which we can obtain a rigorous upper bound on the global optimum. If a good approximation to a feasible point exists, meeting these specifications involves simply one or more iterations of the Krawczyk method or other interval Newton method. This involves only a few interval linear algebra operations but can fail if the constraints are approximately linearly dependent.[49].

Our GLOBSOL software [99, 104] is designed to either satisfy specification 1 or exceed user-set limits on execution time or memory for general problems. Because of this, GLOBSOL's performance does not compare with computer programs designed around the other specifications. However, more recent codes, such as iCOs, still[50] under development for specification 1, promise to be highly competitive.

BARON [227] is successful commercial software for global optimization. It does not claim to rigorously verify the solutions it provides, but it does use interval arithmetic in places to compute bounds on ranges of expressions. Its users can generally expect it to find an approximation to a global optimum and one (of perhaps many) global optimizing points, similar to specification 2.

11.2.7 Additional Centers of Study

Several groups of researchers have been focussing on interval methods in global optimization.

The group headed by Arnold Neumaier at the University of Vienna is doing much scholarly work on global optimization. We have already mentioned Neumaier's relatively early book [167], which remains a valuable reference for the theory underlying interval Newton methods,[51] and the paper [173] that enables easy verification of bounds on the optimum of linear programs. Neumaier's website is a valuable resource for further information on global optimization. Arnold Neumaier, Hermann Schichl, and Waltraud Huyer, along with Oleg Shcherbina and Ferenc Domes, have collaborated in the European Union-wide COCONUT (Continuous CoNstraints—Updating the Technology) project for providing a general computing environment for experimentation with and solution of constraint satisfaction[52] and global optimization problems. One publication of note is a comparison of global optimization software [174]. That publication is one of few attempts to compare existing software in an unbiased way on problems of potentially practical significance and thus provides good guidance. However, the distinctions and goals we have given in section 11.2.6 were not fully made in [174], and further comparison and software development will give additional clarification and increased utility.

[49]This is because the size of the box over which an interval Newton method or the Krawczyk method is contractive is inversely proportional to the maximum condition number of matrices in the interval linear system. The relationship is spelled out, for example, in [97, pp. 219–223].

[50]For the iCOs package, see the web page for our book, www.siam.org/books/ot110.

[51]Interval Newton methods are important for efficiency in solving many global optimization problems.

[52]Constraint satisfaction problems are systems of equality and inequality constraints. They can be viewed as global optimization problems either without an objective function or with a constant objective function. They can be converted into optimization problems through penalty function methods, but that may not be the best approach.

Ferenc Domes, also at the University of Vienna, is developing a MATLAB-based solver GLOPTLAB for rigorous global optimization and for use as a test bed within the COCONUT project.

There has been interest in interval techniques in constraint propagation within the French company ILOG, beginning with a commercial release of Pascal Van Hentenryck's Numerica optimization software in the 1990s [245], incorporation into the ILOG solver, and continuing with efforts by Christian Bliek, in conjunction with the COCONUT project [171]. ILOG has been in a position to provide quality products in this area, since they can interface their CPLEX linear programming solver with techniques for rigorous linear relaxations [28] and the Neumaier–Shcherbina–Jansson technique for providing rigorous bounds on the global optimum [84, 173]. Meanwhile, associates Laurent Granvilliers and Frédéric Benhamou have produced the freely available constraint propagation package [52]. Similarly, Lebbah et al. have studied interval constraint solving in their QUADSOLVER algorithm [130], while Lebbah has been making available his ICOS package for solving systems of constraints.

Researchers in Hungary, led by Tibor Csendes at the University of Szeged, have consistently done high-quality work in verified global optimization over the years. Particularly noteworthy are the successes of this group in concrete applications, exemplified by the advances made with interval-based optimization methods in understanding the problem of circle-packing on the square, reviewed in the book[53] [239].

11.2.8 Summary of Links for Further Study

An excellent summary of available software for global optimization and constraint satisfaction problems is given in Neumaier [170]. See also Ratschek and Rokne [211], Hansen and Walster [57], and Kearfott [97]. An excellent introduction to applied interval analysis, which we also mention in section 11.7, is [86].

A number of interesting websites are also available. See Appendix D.

11.3 Structural Engineering Applications

Rigorous numerical analysis of models involving partial differential equations is particularly challenging. For instance, when a partial differential equation is solved numerically, the rigorous error bounds should reflect not only the roundoff error incurred in the solution process but also the discretization error in transforming the equations (differential or integral) into a linear or nonlinear algebraic system. Furthermore, the motivation for employing interval methods may be that the input data has uncertainties; these uncertainties correspond to intervals whose widths are typically much larger than the widths of intervals corresponding to roundoff error alone. It is more challenging to obtain meaningful results when propagating these larger widths through a solution process for a large linear or nonlinear system. Nonetheless, there have been some notable successes.

Structural engineering problems typically give rise to large, structured matrices when the finite element method is used for the discretization. These matrices are typically "assembled" from smaller matrices corresponding to the individual elements into which the

[53] With collaboration of other European researchers.

structure is partitioned or triangulated. There are uncertainties, such as the length or tensile strength of a particular beam to be used in the construction, that translate into interval values associated with each element. If the elements are assembled in the traditional way to form the system of equations to be solved, and we attempt to use off-the-shelf interval methods (such as the Krawczyk method, preconditioned Gauss–Seidel method, or preconditioned Gaussian elimination, as explained in section 7.4), the resulting solution bounds are usually so wide that they are meaningless. Part of the problem is that dependency is introduced during the assembly process, and there is additional dependency in the system itself. Despite this, there is a sizeable advantage to being able to predict sharp bounds on the behavior of a structural system, corresponding to uncertainties in the properties of the building materials used. For instance, this allows guaranteeing the safety of the structures with less construction cost.

Rafi Muhanna and Robert Mullen have been highly successful in bounding the solutions to structural engineering problems with uncertain parameters. To do so, they have taken advantage of the "structure" of the systems of equations, integrating the assembly and solution process. Examples of this work are [144, 160, 161, 162, 163]. Muhanna has established the Center for Reliable Engineering Computing, in which these and other methods featuring rigorous bounds are studied, and conferences featuring these methods have been held periodically there.

Additional papers in this area include [39] and [172].

11.4 Computer Graphics

Problems in computer graphics typically include finding the intersection points of a surface and a line segment, determining that intersection point of a surface and a line segment closest to one end of the line segment, tracing the intersection curves of two surfaces, finding the intersection points of two curves, determining whether one surface is in "front" of another surface, and the like. Typically, the lines, curves, and surfaces are given as parameterizations of relatively low-order polynomials or piecewise polynomials. This gives rise to low-dimensional systems of low-order polynomial equations, something that can be handled relatively easily with interval computations. Furthermore, it is advantageous to have rigorous bounds on such solutions, since the computations are typically embedded into and carried out repeatedly in relatively sophisticated graphical modeling systems; missing a solution or coming to an incorrect conclusion can result in large problems with the final output, or even cause the system to crash.

Challenges in solving such problems involve, among other things, implementing the solution process to be sufficiently efficient for smooth operation of the system in which it is embedded.

Some references in the area are [233] and [46].

11.5 Computation of Physical Constants

One example of use of interval arithmetic in determining precise values of physical constants is described in [81] and [128]. The authors of these articles are at the University of Wuppertal, where experiments are in progress to more accurately determine Newton's gravitational constant. The gravitational constant is determined through a computation, involving quadrature, that relates experimentally measured quantities to the value of the constant. Holzmann, Lang, and Schütt used verified Gaussian quadrature to determine which of the measured quantities were of critical influence on the error in the computed value of the gravitational constant.

11.6 Other Applications

Interval analysis has found applications in a wide variety of different areas. We mention some of these below, along with a few suitable references, just to whet the reader's appetite:

- computation of complex roots of polynomials and complex eigenvalues [35, 47, 72, 165, 190];
- electrical engineering [31, 51, 114, 212, 230];
- mechanical engineering [115, 162, 256].

11.7 For Further Study

A collection of applications, both proposed and completed in 1996, appears in [103]. An excellent applied introduction, including description of global optimization and parameter estimation techniques in some detail and application to robust control, and robotics, as well as details of implementation for automatic differentiation, implementation using the IEEE arithmetic standard, and use of PROFIL/BIAS [111, 112] and C++ for interval arithmetic applications, is [86].

Appendix A

Sets and Functions

Review of Set Concepts

Set theory occupies the very foundations of mathematics, and its terminology is indispensable in a subject such as ours. We therefore devote a few pages to a brief review of some necessary concepts and notation.

Set Notation

A *set* is a collection or assemblage of objects. The objects are called *elements* or *members* of the set. In this section we use uppercase letters such as $A, B, C, \ldots,$ to denote sets, and lowercase letters such as $a, b, c, \ldots,$ to denote set elements. The notation $a \in A$ means that a is an element of A. The notation $a \notin A$ means that a is not an element of A. We may specify a set by listing its members between braces as, for instance, $A = \{a, b, c\}$. This method works best for sets having finitely many elements but is also acceptable for some infinite sets such as the integers:

$$\{\ldots, -3, -2, -1, 0, 1, 2, 3, \ldots\}.$$

Alternatively, we may use the set-builder notation

$$S = \{x : \mathcal{P}(x)\},$$

which defines S as the set of all elements x such that a proposition $\mathcal{P}(x)$ holds. For instance, if $\mathbb{R}$ is the set of all real numbers and a, b are real numbers with $a \leq b$, then we can define the *closed interval* $[a, b]$ as

$$[a, b] = \{x \in \mathbb{R} : a \leq x \leq b\}. \tag{A.1}$$

The *empty set*, denoted $\emptyset$, is a set with no elements. For example, the set of all odd integers that are multiples of two is empty. Another example is the set $\{x : x \in [1, 2] \cap [3, 4]\}$. (We will review the intersection operation $\cap$ shortly.)

Families of Sets

A *family* (or *class*) is a set whose elements are sets themselves. Let us note that the entities

$$a, \qquad \{a\}, \qquad \{\{a\}\}, \qquad \{a, \{a\}\}, \qquad \{\{a\}, \{b\}\}$$

are all different. The first item is not a set. The second item is a set containing one element (a *singleton set*). The third item is a set containing a set, and so on. The third and fifth items are simple examples of families or collections of sets.

A family $\{A_1, A_2, A_3, \ldots\}$ may be denoted by $\{A_i\}$ and referred to as a *sequence* of sets. For a finite family, we may use notation such as $\{A_i : i = 1, \ldots, n\}$ or simply $\{A_1, A_2, \ldots, A_n\}$.

Subsets and Set Equality

We say that A is a *subset* of B, or that B *contains* A, and write

$$A \subseteq B$$

if every member of A is a member of B. We say that A *equals* B and write $A = B$ if and only if $A \subseteq B$ and $B \subseteq A$. If $A \subseteq B$ and $A \neq B$, then A is a *proper subset* of B and we write

$$A \subset B.$$

The empty set is a subset of any set. Indeed, any statement about members of $\emptyset$ holds *vacuously*, because it is impossible to produce a counterexample or contradiction.

Set Operations

The *union* of two sets A and B, denoted $A \cup B$, is the set of all elements that belong to either A or B or both:

$$A \cup B = \{x : x \in A \text{ or } x \in B\}.$$

More generally, for a family $\{A_i\}$ we have

$$\bigcup_i A_i = \{x : x \in A_i \text{ for at least one } i\}. \tag{A.2}$$

The notation $A_1 \cup \cdots \cup A_n$ may be used for a finite family of n sets.

The *intersection* of two sets A and B, denoted $A \cap B$, is the set of all elements that belong to both A and B. That is,

$$A \cap B = \{x : x \in A \text{ and } x \in B\}.$$

More generally,

$$\bigcap_i A_i = \{x : x \in A_i \text{ for every } i\}. \tag{A.3}$$

We may use the notation $A_1 \cap \cdots \cap A_n$ for the intersection of a finite family. Two sets A and B are said to be *disjoint* if $A \cap B = \emptyset$.

The *complement* of A is the set of all elements that are not members of A:

$$A^c = \{x : x \notin A\}.$$

The *set difference* $A - B$ (sometimes denoted $A \setminus B$) is the set of all elements that belong to A but not to B:

$$A - B = \{x : x \in A \text{ and } x \notin B\}.$$

Relations, Equivalence Relations, and Partial Orderings

Let X and Y be sets. The *Cartesian product* of X and Y, denoted $X \times Y$, is the set of all ordered pairs having first element from X and second element from Y:

$$X \times Y = \{(x, y) : x \in X,\ y \in Y\}.$$

A *relation* R *from* X *to* Y is a subset of $X \times Y$. If $(x, y) \in R$, we write $x\,R\,y$.

Example A.1. Let X and Y be the intervals $[1, 2]$ and $[3, 4]$, respectively. The Cartesian product $X \times Y$ is the set of all ordered pairs (x, y) such that $1 \le x \le 2$ and $3 \le y \le 4$. This is, of course, a unit square in the xy-plane. We could define many different relations R from X to Y. One example would be the set $[1, 2] \times \{3\}$. □

A relation R from X to X is said to be a *relation on* X. Such a relation is called

1. *reflexive* if $x\,R\,x$ for all $x \in X$;
2. *symmetric* if $x\,R\,y$ implies $y\,R\,x$ for all $x, y \in X$;
3. *transitive* if $x\,R\,y$ and $y\,R\,z$ imply $x\,R\,z$ for all $x, y, z \in X$;
4. *antisymmetric* if $x\,R\,y$ and $y\,R\,x$ imply $x = y$ for all $x, y \in X$.

A relation on a set is an *equivalence relation* if and only if it is reflexive, symmetric, and transitive.

Example A.2. The relation of geometric similarity (e.g., between similar triangles) is an equivalence relation. □

Example A.3. In modular arithmetic we write $a \equiv b\ (N)$ whenever $a - b$ is exactly divisible by N. In mod-12 (i.e., "clock") arithmetic, for example, $13 \equiv 1\ (12)$. It is easily verified that equivalence mod-N is an equivalence relation. □

Let R be an equivalence relation on X, and let $x \in X$. The *equivalence class of* x *under* R is the set

$$C_x = \{y \in X : y\,R\,x\}.$$

It can be shown that the family

$$\{C_x : x \in X\}$$

is a *partition* of X, i.e., a collection of subsets of X such that every element of X belongs to one and only one set in the collection.

A relation R on a set X is called a *partial ordering* of X if and only if it is transitive. The relations $<$ and $\le$, for example, are both partial orderings of $\mathbb{R}$.

Functions

A relation f from X to Y is a *function* from X to Y if for every $x \in X$ there is exactly one $y \in Y$ such that $(x, y) \in f$. We write

$$f : X \to Y.$$

The set X is the *domain* of f, and Y is the *codomain* of f. The element

$$y = f(x)$$

is the *image* of x under f, and x is a *preimage* of y under f. The *range* of f is the set

$$\text{Ran } f = \{f(x) : x \in X\}.$$

We sometimes denote the domain of f by Dom f.

Example A.4. The relation given at the end of Example A.1 is a function

$$f : [1, 2] \to [3, 4].$$

It happens to be a constant function whose range contains the single value 3. □

The definition of equality of sets allows us to formulate a condition for the equality of functions. Two functions f and g are equal if and only if Dom f = Dom g and $f(x) = g(x)$ for all $x \in$ Dom f. Hence two functions are equal if and only if their domains are the same and they both map elements from that common domain in precisely the same way.

If $f : X \to Y$ with A a nonempty subset of X, the function $f_A : A \to Y$ such that $f_A(x) = f(x)$ for all $x \in A$ is the *restriction* of f to A.

By an *extension* of a function $f : D \to W$, we mean[54] a function $F : T \to Z$ such that $D \subseteq T$, $W \subseteq Z$, and $F(x) = f(x)$ for all $x \in D$.

A function $f : X \to Y$ is said to be

1. *onto* if Ran $f = Y$;
2. *one-to-one* if $f(x_1) = f(x_2) \implies x_1 = x_2$ whenever $x_1, x_2 \in X$.

If $f : X \to Y$ and $g : Y \to Z$, then the *composition* $g \circ f$ is the function from X to Z such that

$$g \circ f(x) = g(f(x)) \text{ for all } x \in X.$$

Because functions are defined only over their domains, $g \circ f$ can exist only if the range of f is included in the domain of g. Composition is generally not commutative; consider, for example, the two functions $f(x) = x^2$ and $g(x) = x + 1$:

$$f \circ g(x) = (x + 1)^2 \neq g \circ f(x) = x^2 + 1.$$

Composition is associative, however, with

$$h \circ (g \circ f) = (h \circ g) \circ f.$$

[54] In this definition, subset containment permits isomorphisms defined by equivalence relations. In our case, we will be able to write $F([x, x]) = f(x)$ because we have already agreed on the equivalence relation $x \sim [x, x]$.

The *identity function* for a set X is the function $I_X : X \to X$ such that $I_X(x) = x$ for all $x \in X$. A function $f : X \to Y$ is *invertible* if there exists a function $g : Y \to X$ such that $g \circ f = I_X$ and $f \circ g = I_Y$. The function g (if it exists) is unique and is *the inverse* f^{-1} of f. It can be shown that a function is invertible if and only if it is one-to-one and onto.

Set Functions

Let

$$f : X \to Y.$$

If $A \subseteq X$, we define

$$f(A) = \{y \in Y : y = f(x) \text{ for some } x \in A\}. \tag{A.4}$$

This is called the *image* of the set A under f. If $B \subseteq Y$, the *preimage* of B under f is the set

$$f^{-1}(B) = \{x \in X : f(x) \in B\}. \tag{A.5}$$

Example A.5. Let f be defined by $g(x) = x^2$ for all $x \in R$. The image of the interval $[-1, 1]$ under g is the interval $[0, 1]$, and we may write

$$g\,([-1, 1]) = [0, 1].$$

The preimage of the interval $[1, 4]$ under g is the union $[-2, -1] \cup [1, 2]$. □

Note carefully the distinction between a set preimage and an inverse function; the function g of the preceding example is not invertible, but we can still consider preimages of various sets under g.

The reader may wish to verify some simple facts about set functions:

1. $f(\emptyset) = \emptyset$ and $f^{-1}(\emptyset) = \emptyset$;
2. $S_1 \subseteq S_2 \implies f(S_1) \subseteq f(S_2)$;
3. $R_1 \subseteq R_2 \implies f^{-1}(R_1) \subseteq f^{-1}(R_2)$;
4. $R = f(S) \implies S \subseteq f^{-1}(R)$;
5. $S = f^{-1}(R) \implies f(S) \subseteq R$.

In addition, if X and Y are nonempty sets and $f : X \to Y$, then the following statements hold:

1. If $\{X_i\}$ is a family of subsets of X, then
   (a) $f\left(\bigcup_i X_i\right) = \bigcup_i f(X_i)$,
   (b) $f\left(\bigcap_i X_i\right) \subseteq \bigcap_i f(X_i)$.
2. If $\{Y_i\}$ is a family of subsets of Y, then

(a) $f^{-1}\left(\bigcup_i Y_i\right) = \bigcup_i f^{-1}(Y_i)$,

(b) $f^{-1}\left(\bigcap_i Y_i\right) = \bigcap_i f^{-1}(Y_i)$.

Note that if $f\colon X \to Y$ with A and B arbitrary subsets of X and Y, respectively, then it is not true general that $f^{-1}(f(A)) = A$ or $f(f^{-1}(B)) = B$. However, we can see that

$$f^{-1}(f(A)) \supseteq A, \qquad\qquad f(f^{-1}(B)) \subseteq B.$$

We also see that the process of taking preimages preserves set containments, unions, and intersections, while the process of taking images preserves only containments and unions.

Appendix B

Formulary

One-Dimensional Intervals

Interval

$$X = \left[\underline{X}, \overline{X}\right] = \{x \in \mathbb{R} : \underline{X} \le x \le \overline{X}\}$$

Degenerate interval

$$x = [x, x]$$

Union and intersection

$$X \cap Y = \{z : z \in X \text{ and } z \in Y\}$$
$$X \cup Y = \{z : z \in X \text{ or } z \in Y\}$$

$$X \cap Y \neq \emptyset \implies \begin{cases} X \cap Y = \left[\max(\underline{X}, \underline{Y}), \ \min(\overline{X}, \overline{Y})\right], \\ X \cup Y = \left[\min(\underline{X}, \underline{Y}), \ \max(\overline{X}, \overline{Y})\right] \end{cases}$$

Interval hull

$$X \underline{\cup} Y = \left[\min\{\underline{X}, \underline{Y}\}, \ \max\{\overline{X}, \overline{Y}\}\right]$$

$$X \underline{\cup} Y = \begin{cases} = X \cup Y & \text{if } X \cap Y \neq \emptyset, \\ \supset X \cup Y & \text{otherwise.} \end{cases}$$

Width, absolute value, midpoint

$$w(X) = \overline{X} - \underline{X}, \qquad |X| = \max(|\underline{X}|, |\overline{X}|), \qquad m(X) = \tfrac{1}{2}(\underline{X} + \overline{X})$$

Order relations

$$X < Y \iff \overline{X} < \underline{Y}$$
$$X \subseteq Y \iff \underline{Y} \le \underline{X} \text{ and } \overline{X} \le \overline{Y}$$

Definitions of arithmetic operations

$$X \oplus Y = \{x \oplus y : x \in X,\ y \in Y\}, \quad \oplus = + \ - \ \cdot \ /$$

Endpoint formulas for arithmetic operations

$$X + Y = \left[\underline{X} + \underline{Y},\ \overline{X} + \overline{Y}\right]$$
$$X - Y = \left[\underline{X} - \overline{Y},\ \overline{X} - \underline{Y}\right]$$
$$X \cdot Y = [\min S, \max S], \quad S = \{\underline{X}\underline{Y},\ \underline{X}\overline{Y},\ \overline{X}\underline{Y},\ \overline{X}\overline{Y}\}$$
$$X/Y = X \cdot (1/Y), \quad 1/Y = \left[1/\overline{Y},\ 1/\underline{Y}\right], \quad 0 \notin Y$$

Properties of arithmetic operations

$$X + Y = Y + X$$
$$XY = YX$$
$$X + (Y + Z) = (X + Y) + Z$$
$$X(YZ) = (XY)Z$$
$$0 + X = X + 0 = X$$
$$1X = X1 = X$$
$$0X = X0 = 0$$
$$X - X = w(X)[-1, 1]$$
$$X(Y + Z) \subseteq XY + XZ$$
$$x \in \mathbb{R} \implies x(Y + Z) = xY + xZ$$
$$YZ > 0 \implies X(Y + Z) = XY + XZ$$
$$X + Z = Y + Z \implies X = Y$$

Relations between width and absolute value

$$w(aX + bY) = |a|w(X) + |b|w(Y)$$
$$w(XY) \le |X|w(Y) + |Y|w(X)$$
$$w(1/Y) \le |1/Y|^2 w(Y) \text{ if } 0 \notin Y$$

Interval in terms of midpoint and width

$$X = m(X) + \tfrac{1}{2}w(X)[-1, 1]$$

Symmetric intervals

$$X \text{ is symmetric} \iff \underline{X} = -\overline{X}$$

$$X, Y, Z \text{ symmetric} \implies \begin{cases} m(X) = 0, \\ |X| = w(X)/2, \\ X = |X|[-1, 1], \\ X + Y = X - Y = (|X| + |Y|)[-1, 1], \\ XY = |X||Y|[-1, 1], \\ X(Y \pm Z) = XY + XZ = |X|(|Y| + |Z|)[-1, 1] \end{cases}$$

$$Y, Z \text{ symmetric} \implies X(Y + Z) = XY + XZ$$

Interval Vectors

$$X = (X_1, \ldots, X_n), \text{ where the } X_i \text{ are intervals}$$

$$x \in X \text{ if } x_i \in X_i \text{ for each } i$$

$$X \cap Y = (X_1 \cap Y_1, \ldots, X_n \cap Y_n)$$

$$X \subseteq Y \text{ if } X_i \subseteq Y_i \text{ for each } i$$

$$w(X) = \max_i w(X_i)$$

$$m(X) = (m(X_1), \ldots, m(X_n))$$

$$\|X\| = \max_i |X_i|$$

Interval Matrices

$$\|A\| = \max_i \sum_j |A_{ij}|$$

$$w(A) = \max_{i,j} w(A_{ij})$$

$$(m(A))_{ij} = m(A_{ij})$$

Interval Valued Functions

United extension

$$\bar{g}(X) = g(X) = \{g(x) : x \in X\}$$

(The united extension is simply the exact range of g over X.)

Monotonicity

$$\bar{f}(X) = \begin{cases} \left[f(\underline{X}), f(\overline{X})\right], & f \text{ increasing}, \\ \left[f(\overline{X}), f(\underline{X})\right], & f \text{ decreasing} \end{cases}$$

Inclusion isotonicity

$$Y_i \subseteq X_i \text{ for each } i \implies F(Y_1, \ldots, Y_n) \subseteq F(X_1, \ldots, X_n)$$

Fundamental theorem of interval analysis

$$\left.\begin{array}{l} F \text{ an inclusion isotonic} \\ \text{interval extension of } f \end{array}\right\} \implies f(X_1, \ldots, X_n) \subseteq F(X_1, \ldots, X_n)$$

Power function

$$X^n = \{x^n : x \in X\} = \begin{cases} \left[\underline{X}^n, \overline{X}^n\right], & \underline{X} > 0 \text{ or } n \text{ odd (or both)}, \\ \left[\overline{X}^n, \underline{X}^n\right], & \overline{X} < 0 \text{ and } n \text{ even}, \\ [0, |X|^n], & 0 \in X \text{ and } n \text{ even} \end{cases}$$

Sequences

Metric

$$d(X, Y) = \max\{|\underline{X} - \underline{Y}|, |\overline{X} - \overline{Y}|\}$$

$$\begin{aligned} & d(X, Y) = 0 \iff X = Y \\ & d(X, Y) = d(Y, X) \\ & d(X, Y) \le d(X, Z) + d(Z, Y) \\ & d(X + Z, Y + Z) = d(X, Y) \\ & X \subseteq Y \implies d(X, Y) \le w(Y) \\ & d(X, 0) = |X| \end{aligned}$$

Convergence

$$\forall \varepsilon > 0,\ \exists N \text{ such that } n > N \implies d(X_k, X) < \varepsilon$$

Lipschitz condition

$$\exists L \text{ such that } X \subseteq X_0 \implies w(F(X)) \le Lw(X)$$

Uniform subdivision of interval vector

$$X_{i,j} = \left[\underline{X}_i + (j-1)\,w(X_i)/N\,,\ \underline{X}_i + j\,w(X_i)/N\right]$$
$$X_i = \cup_{j=1}^{N} X_{i,j}$$
$$w(X_{i,j}) = w(X_i)/N$$
$$X = \cup_{j_i=1}^{N} (X_{1,j_1}, \ldots, X_{n,j_n})$$
$$w(X_{1,j_1}, \ldots, X_{n,j_n}) = w(X)/N$$

Refinement

$$F_{(N)}(X) = \bigcup_{j_i=1}^{N} F(X_{1,j_1}, \ldots, X_{n,j_n})$$

Excess width

$$w(E(X)) = w(F(X)) - w(f(X))$$

Nested sequence

$$X_{k+1} \subseteq X_k \qquad (k = 0, 1, 2, \ldots)$$

Finite convergence in *K* steps

$$\exists K \text{ such that } k \geq K \implies X_k = X_K$$

Iteration

$$X_{k+1} = F(X_k) \cap X_k \qquad (k = 0, 1, 2, \ldots).$$

One-dimensional Newton's method

$$X^{(k+1)} = X^{(k)} \cap N(X^{(k)}), \qquad N(X) = m(X) - f(m(X))/F'(X)$$

Here, $m(X)$ may be replaced by any $y \in X$, and there may be advantages to doing so in particular instances and for particular k.

Integration

Endpoint formula

$$\int_{[a,b]} F(t)\,dt = \left[\int_{[a,b]} \underline{F}(t)\,dt\,,\ \int_{[a,b]} \overline{F}(t)\,dt\right]$$

Polynomials

a, b have same sign $\Longrightarrow$

$$\int_{[a,b]} (A_0 + \cdots + A_q t^q)\,dt = A_0(b-a) + \cdots + A_q(b^{q+1} - a^{q+1})/(q+1);$$

a, b have opposite signs $\Longrightarrow$

$$\int_{[a,b]} (A_0 + \cdots + A_q t^q)\,dt = T_0 + T_1 + \cdots + T_q, \text{ where}$$

$$T_i = \begin{cases} A_i(b^{i+1} - a^{i+1})/(i+1), & i \text{ even}, \\ \left[(\underline{A}_i b^{i+1} - \overline{A} a^{i+1})/(i+1)\,,\ (\overline{A}_i b^{i+1} - \underline{A} a^{i+1})/(i+1)\right], & i \text{ odd}. \end{cases}$$

Enclosure by Taylor expansion

$$x(t) \in \sum_{k=0}^{N-1} (x)_k (t - t_0)^k + R_N([t_0, t])(t - t_0)^N,$$

where $R_N([t_0, t])$ is an interval enclosure for $x_N([t_0, t])$

$$(x)_0 = x(t_0), \quad (x)_k = \frac{1}{k!}\frac{d^k x}{dt^k}(t_0)$$

Recursions for automatic differentiation

$$(uv)_k = \sum_{j=0}^{k} (u)_j (v)_{k-j}$$

$$(u/v)_k = \left(\frac{1}{v}\right)\left\{(u)_k - \sum_{j=1}^{k} (v)_j (u/v)_{k-j}\right\}$$

$$(u^a)_k = \left(\frac{1}{u}\right)\sum_{j=0}^{k-1}\left(a - \frac{j(a+1)}{k}\right)(u)_{k-j}(u^a)_j$$

$$(e^u)_k = \sum_{j=0}^{k-1}\left(1 - \frac{j}{k}\right)(e^u)_j (u)_{k-j}$$

$$(\ln u)_k = \left(\frac{1}{u}\right)\left\{(u)_k - \sum_{j=1}^{k-1}\left(1 - \frac{j}{k}\right)(u)_j(\ln u)_{k-j}\right\}$$

$$(\sin u)_k = \left(\frac{1}{k}\right)\sum_{j=0}^{k-1}(j+1)(\cos u)_{k-1-j}(u)_{j+1}$$

$$(\cos u)_k = -\left(\frac{1}{k}\right)\sum_{j=0}^{k-1}(j+1)(\sin u)_{k-1-j}(u)_{j+1}.$$

Appendix C

Hints for Selected Exercises

1.1. Using a pair of squares, we find that $2\sqrt{2} \le \pi \le 4$. The reader may wish to try a pair of hexagons as well.

1.2.

$$\frac{F_0 - \Delta F}{m_0 + \Delta m} \le a \le \frac{F_0 + \Delta F}{m_0 - \Delta m}.$$

1.3. For the interval $[3.7, 3.8]$ we have $m = (3.7+3.8)/2 = 3.75$ and $w = 3.8 - 3.7 = 0.1$. Therefore $M \approx 3.75$ kg and $|M - 3.75| \le 0.05$ kg.

1.4. Take natural logs of both sides of $x_{75} = (1 - 10^{-21})^{2^{75}}$ and use a Taylor expansion:

$$\begin{aligned} \ln x_{75} &= 2^{75} \ln(1 - 10^{-21}) \\ &= 2^{75}(-10^{-21} - (1/2)(10^{-42}) - \cdots) \\ &= -37.77893\ldots \end{aligned}$$

and so $x_{75} = 3.9157\ldots \cdot 10^{-17} < 10^{-10}$.

1.5. We can use techniques covered later to prove that we have used enough digits of precision. However, since we are given the correct answer, we can use a variable-precision package to try different precisions, until we find a precision below which we obtain an incorrect answer and above which we obtain a correct answer.

2.1. $X + Y = [3, 13]$ and $X \cup Y = [-2, 7]$.

2.2. $X - Y = [1, 8]$.

2.3. No. The interval expression $X - X$ yields the set of *all* real differences $x - y$ when x and y are permitted to range through $[\underline{X}, \overline{X}]$ *independently*. So, for example, we have $[0, 1] - [0, 1] = [-1, 1]$.

2.4. (a) $[-2, 2]$, (b) $[-12, 6]$, (c) $\left[-\frac{2}{3}, -\frac{1}{5}\right]$, (d) $\left[-\frac{1}{5}, \frac{2}{5}\right]$.

2.5. The required inner products are given by

$$p = (\tfrac{1}{\sqrt{5}}, \tfrac{2}{\sqrt{5}}, 0) \cdot (f, 6, -7) = \tfrac{1}{\sqrt{5}}(f + 12).$$

Replacing f by the interval $F = [1, 3]$, we find that p is enclosed in the interval

$$P = \frac{1}{\sqrt{5}}([1,3] + 12) = \frac{1}{\sqrt{5}}[13, 15] = [5.8137767\ldots, 6.7082039\ldots].$$

Later we will introduce a procedure called *outward rounding*. In the present instance, outward rounding to one place gives us $p \in [5.8, 6.8]$.

3.1. To retain the third place, we truncate at place 4 for the left endpoint and round upward at place 3 for the right endpoint: $[1.234, 1.457]$.

3.2. Direct computation shows that the volume $v = wlh$ lies within in the interval $V = [42599.432, 44435.1]$, hence $v = 43517.266 \pm 917.834$. If desired, we could use outward rounding to retain containment with numbers having only one decimal place: $v \in [42599.4, 44435.1]$.

3.4. For Exercise 3.2, we can enter

```
W = infsup(7.1,7.3);
L = infsup(14.8,15);
H = infsup(405.4,405.8);
V = W*L*H
```

4.3. *Proof of* (4.4).

$$\begin{aligned}
\xi \in X(Y+Z) &\implies \xi = x\zeta \text{ for some } x \in X \text{ and } \zeta \in Y+Z \\
&\implies \xi = x(y+z) \text{ for some } x \in X,\ y \in Y, \text{ and } z \in Z \\
&\implies \xi = xy + xz \text{ for some pair of numbers } x \in X,\ y \in Y \\
&\qquad\qquad \text{and some pair of numbers } x \in X,\ z \in Z \\
&\implies \xi \in XY + XZ.
\end{aligned}$$

Proof of (4.6). By (4.4), it suffices to show that $XY + XZ \subseteq X(Y+Z)$. Let $a \in XY + XZ$. Then

$$a = x_1 y + x_2 z \text{ for some } x_1, x_2 \in X,\ y \in Y,\ z \in Z.$$

By hypothesis $yz \geq 0$, so we have $y + z = 0$ iff $y = z = 0$. If $y = z = 0$, then a is clearly in $X(Y+Z)$. Otherwise $y + z \neq 0$, and choosing

$$x = x_1 \frac{y}{y+z} + x_2 \frac{z}{y+z} \in X,$$

we have $a = x(y+z) \in X(Y+Z)$.

4.4.

$$\begin{aligned}
X + Z = Y + Z &\implies [\underline{X} + \underline{Z}, \overline{X} + \overline{Z}] = [\underline{Y} + \underline{Z}, \overline{Y} + \overline{Z}] \\
&\implies \underline{X} + \underline{Z} = \underline{Y} + \underline{Z} \text{ and } \overline{X} + \overline{Z} = \overline{Y} + \overline{Z} \\
&\implies \underline{X} = \underline{Y} \text{ and } \overline{X} = \overline{Y} \\
&\implies X = Y.
\end{aligned}$$

4.5. Take $Z = [0, 2]$, $X = [0, 1]$, and $Y = [1, 1]$.

4.6.

$$X = [\underline{X}, \overline{X}] = \frac{1}{2}(\underline{X} + \overline{X}) + \frac{1}{2}(\overline{X} - \underline{X})[-1, 1].$$

5.6. The command `setround(-1)` sets the rounding to "round down," the command `setround(1)` sets the rounding to "round up," while the command `setround(0)` sets the rounding to "round to the nearest machine number." There is also some complication in this routine, so it can handle arrays and complex interval arguments, and so it exits with the machine in the same rounding mode as when the routine was entered. However, try to identify the part where the lower and upper bounds of the interval value `y` are set.

5.8. If F is an interval extension of f such that $F(x) = f(x)$ for real x, then $F_1(X) = F(X) + X - X$ defines a different interval extension of f.

5.9. This can be done in various ways, plotting on a single window or in multiple windows. For example, plotting f in its own window could be done with the following sequence:

```
>> x = linspace(0.999,1.001);
>> y = x.^6-6*x.^5.+15.*x.^4-20.*x.^3.+15.*x.^2-6.*x+1;
>> plot(x,y)
```

5.10. Setting $p'(x) = -5 + x^2 = 0$, we obtain $x = \pm\sqrt{5}$. Now $\sqrt{5}$ is in $[2, 3]$ and furthermore, $p''(x) = 2x > 0$ for x in $[2, 3]$, so $\sqrt{5}$ is a local minimum of $p(x)$ for x in $[2, 3]$. Since it is the only place in $[2, 3]$ where $p'(x) = 0$, it remains only to evaluate $p(x)$ at the endpoints 2 and 3. Since $p(2) = -6.3333\ldots$ and $p(3) = -5$, we obtain (5.28).

5.11. $F([0, 2]) = \left[\frac{1}{2}, \frac{3}{2}\right]$, but $F([0, 1]) = \left[\frac{1}{4}, \frac{3}{4}\right] \not\subseteq F([0, 2])$.

5.13. Since $f'(x) = x^2 - 2$, we have $f'(x) = 0$ at $x = \sqrt{2} \in [0, 3]$. So $f(x)$ has a minimum value at $x = \sqrt{2}$, namely, $f(\sqrt{2}) = -\frac{4}{3}\sqrt{2}$. Now the maximum value is easily seen to be $f(3) = 3$. Therefore $f([0, 3]) = [-\frac{4}{3}\sqrt{2}, 3]$. With further calculation, we find that $f([0, 3]) \subseteq [-1.88562, 3]$.

Without using calculus, we can find intervals enclosing the range of values found above. First, $f([0, 3]) \subseteq \frac{1}{3}[0, 3]^3 - 2[0, 3] = [-6, 9]$. The form $f(x) = x(\frac{1}{3}x^2 - 2)$, equivalent in real arithmetic, gives $f([0, 3]) \subseteq [0, 3](\frac{1}{3}[0, 3]^2 - 2) = [0, 3][-2, 1] = [-6, 3]$.

6.2. (a) We say that $F(X)$ is continuous at X_0 if for every $\varepsilon > 0$ there is a positive number $\delta = \delta(\varepsilon)$ such that $d(F(X), F(X_0)) < \varepsilon$ whenever $d(X, X_0) < \delta$. Here, again, d is the interval metric given by (6.3).

6.3. Let $A = [\underline{A}, \overline{A}]$, $A' = [\underline{A}', \overline{A}']$, $B = [\underline{B}, \overline{B}]$, $B' = [\underline{B}', \overline{B}']$ and write

$$\begin{aligned} d(A+B, A'+B') &= d([\underline{A}+\underline{B}, \overline{A}+\overline{B}], [\underline{A}'+\underline{B}', \overline{A}'+\overline{B}']) \\ &= \max\{|(\underline{A}+\underline{B})-(\underline{A}'+\underline{B}')|, |(\overline{A}+\overline{B})-(\overline{A}'+\overline{B}')|\} \\ &= \max\{|\underline{A}-\underline{A}'+\underline{B}-\underline{B}'|, |\overline{A}-\overline{A}'+\overline{B}-\overline{B}'|\} \\ &\le \max\{|\underline{A}-\underline{A}'|+|\underline{B}-\underline{B}'|, |\overline{A}-\overline{A}'|+|\overline{B}-\overline{B}'|\} \\ &\le \max\{|\underline{A}-\underline{A}'|, |\overline{A}-\overline{A}'|\} + \max\{|\underline{B}-\underline{B}'|, |\overline{B}-\overline{B}'|\} \\ &= d(A, A') + d(B, B'). \end{aligned}$$

Hence given any $\varepsilon > 0$ we can choose $\delta = \varepsilon/2$ and have $d(A+B, A'+B') < \varepsilon$ whenever both $d(A, A')$ and $d(B, B')$ are less than δ. In fact, δ is dependent on ε only (and not on the intervals A, A', B, B') so the continuity is uniform.

6.10. From

$$H(Y) = -2 + \tfrac{1}{3}\left((Y+c)^2 + (Y+c)c + c^2\right)$$

we obtain $F_c([1.2, 1.6]) \subseteq [-1.95, -1.83]$. The exact minimum value for $x \in [1.2, 1.6]$ is at $x = \sqrt{2}$ and is $f(\sqrt{2}) = -1.8856\ldots$. So the exact range of values is

$$[-1.8856\ldots, -1.824\ldots]$$

because $\max(f(1,2), f(1.6)) = -1.824\ldots$.

6.11. Since $f'(x) = x^2 - 2$, we take $DF(X) = X^2 - 2$. Then $DF([1.2, 1.6]) = [-0.56, 0.56]$ and

$$\begin{aligned} F_{mv}([1.2, 1.6]) &= f(1.4) + [-0.56, 0.56][-0.2, 0.2] \\ &= -1.885333\ldots + [-0.112, 0.112] \\ &\subseteq [-2.01, -1.77]. \end{aligned}$$

In this example (as in most others) F_c is narrower.

7.5. We have

$$\begin{aligned} (R_1 + R_2)I_1 - R_2 I_2 &= V_1, \\ -R_2 I_1 + (R_2 + R_3)I_2 &= -V_2, \end{aligned}$$

with $V_1 = 10$, $V_2 = 5$, and interval coefficients determined by $R_1 = R_2 = R_3 = 1000 \pm 10\%$. So $R_1 = R_2 = R_3 = [900, 1100]$. We can apply the Gaussian elimination process shown in the text. We first notice that the multiplier a_{21}/a_{11} becomes $-R_2/(R_1 + R_2)$ which, as we have seen in an earlier chapter, is better written as $-1/(1 + R_1/R_2)$ for interval computation. In fact, with the given data $R_1 = R_2 = R_3 = [900, 1100]$, the first expression yields

$$-[900, 1100]/([900, 1100] + [900, 1100]) = -[\tfrac{9}{22}, \tfrac{11}{18}] = -[0.409\ldots, 0.611\ldots],$$

whereas the second one yields the narrower interval enclosure

$$-1/(1 + [900, 1100]/[900, 1100]) = -[\tfrac{9}{20}, \tfrac{11}{20}] = -[0.45, 0.55].$$

Using this interval multiplier, the Gaussian formulas, evaluated in IA at the third or fourth nonzero digit, yield the results

$$\begin{aligned}&([1800, 2200] - (-[0.45, 0.55])(-[900, 1100]))I_2 = -5 - (-[0.45, 0.55])10,\\&[1800, 2200]I_1 = 10 - (-[900, 1100])I_2,\end{aligned}$$

so

$$[1195, 1795]I_2 = -5 + [4.5, 5.5] = [-0.5, 0.5]$$

and

$$I_2 = [1/1795, 1/1195][-0.5, 0.5] = [-0.000419, 0.000419]$$

and

$$\begin{aligned}I_1 &= [1/2200, 1/1800](10 + [900, 1100][-0.000419, 0.000419])\\&= [1/2200, 1/1800](10 + [-0.4609, 0.4609])\\&= [1/2200, 1/1800][9.5291, 10.4609].\end{aligned}$$

Our result is that the interval vector (I_1, I_2) encloses the set of all real vector solution points for any real number choices of coefficients from the given interval coefficients, where

$$I_1 = [0.00433, 0.00582] \quad \text{and} \quad I_2 = [-0.000419, 0.000419].$$

8.3.

$$m(X^{(0)}) - \frac{[m(X^{(0)})]^2 - 2}{2X^{(0)}} = 5 - \frac{25 - 2}{2[4, 6]} = [\tfrac{17}{8}, \tfrac{37}{12}].$$

8.6. $(-\infty, -\frac{1}{24}] \cup [\frac{1}{3}, \infty)$.

8.7. We seek x such that $x^3 = 5x + 1$. To find a good starting interval, examine the graphs of $y = x^3$ and $y = 5x + 1$.

8.8. First note that an object with specific gravity < 1 will float. The given polynomial has three real roots in the interval $[-1, 3]$. However, only a root lying in $[0, 2]$ makes sense as an answer. We find $h = 1.22 \pm 0.01$.

8.10. In Case 1, although the numerator and denominator of the quotient $1/\overline{X}$ are both real numbers, we convert them to intervals to obtain a *mathematically rigorous* lower bound. (It is necessary that the computed result contain the exact mathematical result.) Thus, $1/\overline{X}$ is computed with interval arithmetic, and the lower bound of that result is used to form the lower bound of the semi-infinite interval. A similar technique is used for Case 2 and Case 3. Also, the Cset model used in this function has underlying set of numbers the finite reals (i.e., it does not include $-\infty$ and ∞ as numbers). If the extended reals were used as the underlying model, then in Case 1 we would have `Y2 = infsup(-Inf,-Inf)`, in Case 3 we would have `Y2 = infsup(Inf,Inf)`, and in Case 4 we would have `Y1 = infsup(-Inf,-Inf)` and `Y2 = infsup(Inf,Inf)`. Caution should thus be exercised when using `xreciprocal` in either case, since INTLAB follows the finite version of Cset theory in some cases, if one interprets `[NaN,NaN]` as the empty set, but INTLAB does not follow it consistently: Try computing `infsup(0,0)/infsup(-1,1)`, `infsup(0,0)/infsup(0,0)`, etc. Also see Note 3 on p. 114.

8.11. In the given equation $f(I) = 0$, the function $f(I)$ can be simplified:

$$\ln(10^9 I) = \ln(10^9) + \ln(I),$$

so

$$f(I) = 0.8922390 - I - 0.0052\ln(I).$$

$f(I)$ is clearly negative for $I > 1$ and positive for $I < 0.892239$. Thus, there is exactly one real positive zero in [0.8922390, 1.0]. We have

$$f'(I) = -1 - 0.0052/I$$

so starting with $X = [0.892239, 1.0]$, we can compute an interval Newton iteration as follows:

$$\begin{aligned}
m(X) &= 0.9461195,\\
f(m(X)) &= 0.892239 - 0.9461195 - 0.0052\ln(0.9461195)\\
&= -0.053592491,\\
f'(X) &= f'([0.892239, 1.0])\\
&= -1 - 0.0052/[0.892239, 1.0]\\
&= [-1.005828036, -1.0052],\\
N(X) &= 0.9461195 - \{-0.053592491/[-1.005828036, -1.0052]\}\\
&= [0.8928042, 0.8928376].
\end{aligned}$$

Intersecting that with $X = [0.892239, 1.0]$ does not change it, because $N(X)$ is already contained in X. In fact it contains the unique solution to the problem and is already an interval of width about 3×10^{-5}.

8.12. Aberth's RANGE yields the following solutions: (1) for the search interval [1.25, 1.35], the solution $\theta_i = 1.30958$ radians (75.0334°); (2) for the search interval [0.73, 0.9], the solution $\theta_i = 0.76551$ radians (43.8605°).

8.13. As can be seen graphically, there are two solutions

$$X = \pm\begin{pmatrix}0.7071\cdots\\0.7071\cdots\end{pmatrix}.$$

8.14. $I = 0.94641$, $V = 1.071715$.

8.15. There are three critical points: to five places they are (0.06264, 0.75824), (0.28537, 0.60975), and (0.88443, 0.21038).

To see that the starting box $X_0 = ([0, 1], [0, 1])$ is large enough, superpose two plots of x_2 vs. x_1: one from the first equation (involving the fifth-degree polynomial) and one from the second equation (straight line). It should be obvious that there are three intersection points, all having $x_1 \in [0, 1]$. Then return to the straight line equation to see that these intersections all have $x_2 \in [0, 1]$ as well.

8.20. First, your results may differ, depending upon how INTLAB is set to display intervals. Observe the following dialogue:

```
>> intvalinit('DisplayMidrad')
>> sqrt(infsup(-1,1))
intval ans = < 0.0000 +  0.0000i , 1.0000 >
>> intvalinit('Display__')
>> sqrt(infsup(-1,1))
intval ans =   0_.____ + 0_.____i
>> intvalinit('DisplayInfsup')
>> sqrt(infsup(-1,1))
intval ans = [ -1.0000 - 1.0000i , 1.0000 + 1.0000i ]
```

In the first case, the result is represented as a disk of radius 1 in the complex plane, while, in the last case, the result is represented as a rectangle in the complex plane, whose lower left corner is $-1-i$ and whose upper right corner is $1+i$. Thus, INTLAB interprets `sqrt` as being complex-valued for negative values of its argument. In fact, INTLAB uses the principal value of the square root; you can see this when `sqrt(infsup(-1,-1))` returns a small region about i. This interpretation of the square root is appropriate in some contexts, but not in others, such as in constraint propagation with real variables or in interval Newton methods dealing with real variables.

9.1. For k odd we have

$$At^k = \left[\underline{At^k}, \overline{At^k}\right] = \begin{cases} \left[t^k\underline{A}, t^k\overline{A}\right], & t > 0, \\ \left[t^k\overline{A}, t^k\underline{A}\right], & t < 0. \end{cases}$$

Therefore

$$\begin{aligned}
\int_{[a,b]} At^k\,dt &= \int_{[a,0]} At^k\,dt + \int_{[0,b]} At^k\,dt \\
&= \left[\int_{[a,0]} \underline{At^k}\,dt, \int_{[a,0]} \overline{At^k}\,dt\right] + \left[\int_{[0,b]} \underline{At^k}\,dt, \int_{[0,b]} \overline{At^k}\,dt\right] \\
&= \left[\int_{[a,0]} \overline{A}t^k\,dt, \int_{[a,0]} \underline{A}t^k\,dt\right] + \left[\int_{[0,b]} \underline{A}t^k\,dt, \int_{[0,b]} \overline{A}t^k\,dt\right] \\
&= \left[-\frac{a^{k+1}}{k+1}\overline{A}, -\frac{a^{k+1}}{k+1}\underline{A}\right] + \left[\frac{b^{k+1}}{k+1}\underline{A}, \frac{b^{k+1}}{k+1}\overline{A}\right] \\
&= \left[\frac{b^{k+1}\underline{A} - a^{k+1}\overline{A}}{k+1}, \frac{b^{k+1}\overline{A} - a^{k+1}\underline{A}}{k+1}\right].
\end{aligned}$$

Note: The result is equivalent to

$$\{b^{k+1}/(k+1)\}A - \{a^{k+1}/(k+1)\}A, \qquad (*)$$

but this is *not* the same as $\{b^{k+1}/(k+1) - a^{k+1}/(k+1)\}A$. For example, if $k = 3$ and $A = [1, 2]$, and $[a, b] = [-1, 1]$, then (*) is $(1/4)[1, 2] - (1/4)[1, 2]$, which is $[-1/4, 1/4]$—not $[0, 0]$.

9.3. To 10 places, Aberth's RANGE finds (a) 6.74300141925; (b) 0.6314179219.

9.4. From (9.12) and $F(t) \subseteq G(t)$, we have $\underline{G} \le \underline{F}$ and $\overline{F} \le \overline{G}$ so (9.13) follows.

9.6. The first two formulas are an exercise in elementary calculus. The third is essentially the Leibnitz formula

$$\frac{d^k(uv)}{dx^k} = \sum_{j=0}^{k} \binom{k}{j} \frac{d^j u}{dx^j} \frac{d^{k-j} v}{dx^{k-j}},$$

where the binomial coefficient

$$\binom{k}{j} = \frac{k!}{j!(k-j)!}$$

can be broken up into its three factors to obtain (dividing both sides of the Leibnitz formula by $k!$)

$$\frac{1}{k!}\frac{d^k(uv)}{dx^k} = \sum_{j=0}^{k} \left(\frac{1}{j!}\frac{d^j u}{dx^j}\right)\left(\frac{1}{(k-j)!}\frac{d^{k-j} v}{dx^{k-j}}\right) = \sum_{j=0}^{k} (u)_j (v)_{k-j}.$$

9.7. (b) With $u(t) = t$ we have $(u)_1 = 1$ and $(u)_i = 0$ for $i > 1$. So only the $j = 0$ terms contribute in the summations for the sine and cosine.

9.10. The integrand of (9.34) is $f(t) = 1/t$. We have

$$(f)_K(t) = \frac{1}{K!}\frac{d^K}{dt^K}(1/t) = \frac{(-1)^K}{t^{K+1}},$$

so

$$(f)_K(X_i) = \frac{(-1)^K}{X_i^{K+1}},$$

where X_i is a subinterval of the integration interval $[1, 2]$. Write $X_i = \left[\underline{X}_i, \overline{X}_i\right]$ and carry out the interval arithmetic:

$$(f)_K(X_i) = \frac{(-1)^K}{[\underline{X}_i, \overline{X}_i]^{K+1}} = \frac{(-1)^K}{[\underline{X}_i^{K+1}, \overline{X}_i^{K+1}]} = (-1)^K \left[\frac{1}{\overline{X}_i^{K+1}}, \frac{1}{\underline{X}_i^{K+1}}\right].$$

For K even, the left side of (9.40) is

$$w((f)_K(X_i)) = \frac{1}{\underline{X}_i^{K+1}} - \frac{1}{\overline{X}_i^{K+1}}.$$

From this, since $1 \le \underline{X}_i < \overline{X}_i \le 2$, it follows that

$$w((f)_K(X_i)) \le (K+1)(\overline{X}_i - \underline{X}_i).$$

Just write

$$\frac{1}{\underline{X}_i^{K+1}} - \frac{1}{\overline{X}_i^{K+1}} = \frac{\overline{X}_i^{K+1} - \underline{X}_i^{K+1}}{\underline{X}_i^{K+1}\overline{X}_i^{K+1}} = (\overline{X}_i - \underline{X}_i)[\text{sum of } K+1 \text{ terms}],$$

each of them ≤ 1.

9.11.

$$\int_0^1 \int_0^{1-x_1} \int_0^{1-x_1-x_2} (x_1^2 + x_2^2 + x_3^2) e^{-x_1 x_2 x_3} \, dx_3 \, dx_2 \, dx_1 = 0.04955715430\tilde{}.$$

10.3. (a) We can take $Y^{(0)} = [1, b]$ for $b > 1$ and find that

$$P(Y^{(0)})(t) = 1 + \int_0^t [1, b^2] \, ds = 1 + [1, b^2]t \subseteq [1, b] \text{ for } t \in [0, (b-1)/b^2].$$

The maximum value that $(b-1)/b^2$ can have for any $b \geq 1$ is $1/4$ when $b = 2$. The exact solution is $y(t) = 1/(1-t)$, which blows up as $t \to 1$. (b) The exact solution is $y(t) = e^{-t}$.

Appendix D

Internet Resources

Interval analysis is a rapidly evolving area with an ever-expanding set of applications. Since the subject is inherently computational, it is not surprising that the Internet has played an essential role in its growth. The World Wide Web is replete with information on interval analysis—both theoretical and applied—and the reader is strongly urged to explore as much of this material as may interest him or her. The purpose of this appendix is merely to suggest two starting points.

Interval Computations Site

The Interval Computations site

    http://www.cs.utep.edu/interval-comp/

serves as a main hub for the interval analysis community and is strongly recommended. It enumerates various programming languages for interval analysis, provides links to interval and related software packages, and lists homepages of some active interval researchers. Interval Computations originated in 1995 during an international workshop held at El Paso, Texas. According to Vladik Kreinovich, comaintainer of the site, its goal is to disseminate information and facilitate collaboration between researchers both inside and outside the interval community. There are two mirror sites: one at the University of Texas at El Paso and one in Spain.

Web Page for This Book

The authors will maintain a web page at

    http://www.siam.org/books/ot110

Here they will endeavor to offer current links to the computational tools mentioned in the book (such as INTLAB, RANGE, and GlobSol), as well as other resources of interest to the reader.

Appendix E

INTLAB Commands and Functions

Here, we list some INTLAB commands and functions. Many MATLAB functions and commands are overloaded in INTLAB, and not all of these may be listed here. MATLAB's help feature may provide information about whether a particular function can be applied to intervals. For example, if you are interested in an interval extension of the sine function, you can type `help sin` from a MATLAB command window. You will then see basic MATLAB help for `sin`, along with a listing, in blue, of links to directories of definitions for other data types. In particular, you will see `help intval/sin.m`. This indicates that there is support for an interval version of `sin`. Click on this blue link for further information.

With INTLAB installed, it is useful to go through the INTLAB part of MATLAB's `Help` → `Demos` menu. Finally, information and various examples are available at the INTLAB web page.

We have not included functions from INTLAB's accurate sum, gradient, Hessian, slope, and variable precision arithmetic toolboxes in this appendix. The information here applies to version 5.5 of INTLAB (and, presumably, later versions).

Some Basic Commands

Here we let X denote an interval, and a, b, c, and d real numbers.

`intvalinit('DisplayInfsup'):` inf-sup default display

`intvalinit('DisplayMidrad'):` midpoint-radius default display

`intvalinit('Display_'):` uncertainty representation default display

`X = infsup(a,b):` defines X as $[a, b]$

`X = midrad(a,b):` defines X as $[c, d] = [a - b, a + b]$

`+, -, *, /, \, ^, ' :` interval arithmetic operations, as defined in MATLAB. As in MATLAB, these can be applied to matrices.

`== , >=, <= , >, <, ~= :` logical operators, as in MATLAB, but with interval interpretations

Some Special Interval Functions

These are functions special to the interval computations. Many can be applied when X and Y are interval vectors or matrices.

diam(X): width of X

hull(X,Y): interval hull of X and Y

in(X,Y): returns 1 if and only if X is in Y (X may be real or interval, and X and Y may be matrices)

in0(X,Y): returns 1 if and only if X is in the interior of Y (X may be real or interval, and X and Y may be matrices)

inf(X): lower bound on X

intersect(X,Y): intersection of X and Y

inv(A): interval inverse of the matrix A, if it can be computed

isempty(X): returns 1 if and only if X is the empty interval

isintval(X): returns 1 if and only if X is an interval data type

mag(X): magnitude of X (applied componentwise for arrays)

mid(A): midpoint of A

mig(X): mignitude of X

rad(X): radius of the interval X (half of `diam(X)`)

sup(X): upper bound on X

Some Standard Functions Implemented in INTLAB

These functions return interval bounds on the ranges of the corresponding real-valued functions. The bounds are generally good approximations to the narrowest possible bounds.

abs	asin	coth	sin
acos	asinh	csc	sinh
acosh	atan	exp	sqrt
acot	atanh	log	sum
acoth	cos	log10	tan
acsc	cosh	log2	tanh
asec	cot	sec	

INTLAB's Plotting Facilities

INTLAB has overloaded MATLAB's `plot`, `semilogx`, and `semilogy` routines and also has special routines `plotintval` and `plotlinsol`.

`plot(x,Y),plot(X,y),plot(X,Y)`: For the abscissa vector x real and the corresponding ordinate vector Y interval, the INTLAB version plots the lower bounds on Y and the upper bounds on Y as curves, shading in between. For both X and Y intervals, `plot(X,Y)` plots shaded boxes for each (X_i, Y_i) pair. If X is interval and y is real, `plot(X,y)` plots a horizontal line for each (X, y) pair.

`semilogx(X,Y), semilogy(X,Y), loglog(X,Y)`: Overall, these behave as `plot` for one or both of abscissa X and ordinate Y interval. However, when x is real and Y is interval, the area between the lower bound curve and upper bound curve is not shaded, as it is for `plot`. As in the versions that come with MATLAB, `semilogx`, `semilogy`, and `loglog` plot the x-axis, y-axis, and both axes on a logarithmic scale, respectively.

`plotintval(X)`: If X is an n-dimensional complex interval vector, `plotintval` plots each component as a circle in the complex plane. If X is a two-dimensional array of real intervals with dimensions `(2,N)` representing lower bounds and upper bounds of an N-dimensional interval vector, `plotintval` plots each interval vector $[X_{1,i}, X_{2,i}]$ as a box in the xy-plane. Neither the circles nor the boxes are shaded, but their outlines are plotted in colors that cycle.

`plotlinsol(A,B)`: This routine plots the solution set to an interval linear system $AX = B$ in two unknowns. (Such a solution set is illustrated in Figure 7.1 on p. 89.) The actual solution set is shaded, and the entire solution set is enclosed in its interval hull (the narrowest possible interval vector enclosing the solution set).

Some Other MATLAB Functions Overloaded in INTLAB

The following is a partial list of additional MATLAB functions that can be applied to interval data types:

`diag`	`isnan`	`norm`	`spdiags`
`dim`	`isreal`	`real`	`spones`
`disp2str`	`issparse`	`repmat`	`spy`
`display`	`iszero`	`reshape`	`trace`
`find`	`length`	`numel`	`transpose`
`full`	`imag`	`relerr`	`tril`
`isfinite`	`logical`	`size`	`triu`
`isinf`	`nonzeros`	`sparse`	

Particularly noteworthy is INTLAB's support for sparse matrices.

INTLAB's Matrix-Vector Arithmetic

INTLAB uses midpoint-radius representation of intervals in matrix-vector multiplication in such a way that the rounding mode (round down or round up) need not be switched after each individual interval-interval multiplication. However, a consequence is that the result of the individual multiplications is not sharp. The results are sharp when the intervals are symmetric about 0, e.g., $[-3, 3]$, and such intervals occur in the Krawczyk method in $E = I - YA$ when Y is a good approximation to $m(A)^{-1}$ (see (7.7) on p. 91) or in $(X - y)$ when $y - m(X)$ (see (8.13) on p. 116). In other cases, the results can be far from exact. For example, to check Example 7.2 on p. 86, we see the following:

```
>> intvalinit('FastIVmult')
===> Fast interval matrix multiplication in use (maximum
         overestimation factor 1.5 in radius)
>> A = [infsup(1,2) infsup(3,4)]
intval A =
[ 1.0000 , 2.0000 ] [ 3.0000 , 4.0000 ]
>> B = [infsup(5,6) infsup(7,8); infsup(9,10), infsup(11,12)]
intval B =
[ 5.0000 , 6.0000 ] [ 7.0000 , 8.0000 ]
[ 9.0000 , 10.0000 ] [ 11.0000 , 12.0000 ]
>> A*B
intval ans =
[ 31.0000 , 52.0000 ] [ 39.0000 , 64.0000 ]
intvalinit('SharpIVmult')
===> Slow but sharp interval matrix multiplication in use
>> A*B
intval ans =
[ 32.0000 , 52.0000 ] [ 40.0000 , 64.0000 ]
```

The `startintlab.m` file distributed with INTLAB may specify fast matrix-vector multiplication, so you will need to issue `intvalinit('SharpIVmult')` to get results that are almost exact. Nonetheless, in contexts such as the Krawczyk method and for very large systems that are point matrices or almost point matrices, the fast multiplication is appropriate. INTLAB uses the fast multiplication internally for various computations.

Some Other INTLAB Routines

These implement some other useful features only in INTLAB, as well as certain less fundamental algorithms in interval computations:

`circulant`, `gershgorin`, `lssresidual`, `toeplitz`.

Users of INTLAB have contributed numerous other routines besides the ones listed here. For instance, Jiri Rohn has made publicly available an extensive library of routines for the analysis of interval linear systems.

References

[1] Oliver Aberth. *Precise Numerical Methods Using C++*. Academic Press, New York, 1998.

[2] Oliver Aberth. *Introduction to Precise Numerical Methods*. Academic Press, New York, 2007.

[3] Azmy S. Ackleh, Edward J. Allen, Ralph Baker Kearfott, and Padmanabhan Seshaiyer, editors. *Classical and Modern Numerical Analysis: Theory, Methods, and Practice*. Taylor and Francis, Boca Raton, FL, 2009.

[4] Götz Alefeld. *Intervallrechnung über den komplexen Zahlen und einige Anwendungen*. Dissertation, Universität Karlsruhe, Germany, 1968.

[5] Götz Alefeld. Bemerkungen zur Einschließung der Lösung eines linearen Gleichungssystems. *Z. Angew. Math. Mech.*, 50:T33–T35, 1970.

[6] Götz Alefeld and Nikolaos Apostolatos. Praktische Anwendung von Abschätzunsformeln bei Iterationsverfahren. *Z. Angew. Math. Mech.*, 48:T46–T49, 1968.

[7] Götz Alefeld and Jürgen Herzberger. Über de Verbesserung von Schranken für die Lösung bei linearen Gleichungssystemen. *Angew. Informatik (Elektron. Datenverarbeitung)*, 13:107–112, 1971.

[8] Götz Alefeld and Jürgen Herzberger. Zur Invertierung linearer und beschränkter Operatoren. *Math. Zeitschr.*, 120:309–317, 1971.

[9] Götz Alefeld and Jürgen Herzberger. Einschließungsverfahren zur Berechnung des inversen Operators. *Z. Angew. Math. Mech.*, 52:T197–T198, 1972.

[10] Götz Alefeld and Jürgen Herzberger. Nullstelleneinschließung mit dem Newton-Verfahren ohne Invertierung von Intervallmatrizen. *Numer. Math.*, 19(1):56–64, Feb. 1972.

[11] Götz Alefeld and Jürgen Herzberger. Ein Verfahren zur monotonen Einschließung von Lösungen nichtlinearer Gleichungssysteme. *Z. Angew. Math. Mech.*, 53:T176–T177, 1973.

[12] Götz Alefeld and Jürgen Herzberger. *Einführung in die Intervallrechnung*. Springer-Verlag, Berlin, 1974.

[13] Götz Alefeld and Jürgen Herzberger. *Introduction to Interval Computations*. Academic Press, New York, 1983.

[14] Götz Alefeld, Jürgen Herzberger, and Otto Mayer. Über neuere Gesichtspunkte beim numerischen Rechnen. *Math. Naturw. Unterricht*, 24:458–467, 1971.

[15] Nikolaos Apostolatos and Ulrich W. Kulisch. Approximation der erweiterten Intervallarithmetik durch die einfache Maschinenintervallarithmetik. *Computing*, 2:181–194, 1967.

[16] Nikolaos Apostolatos and Ulrich W. Kulisch. Grundlagen einer Maschinenintervallarithmetik. *Computing*, 2:89–104, 1967.

[17] Nikolaos Apostolatos and Ulrich W. Kulisch. Grundzüge einer Intervallrechnung für Matrizen und einige Anwendungen. *Elektronische Rechenanlagen*, 10:73–83, 1968.

[18] Nikolaos Apostolatos, Ulrich W. Kulisch, Rudolf Krawczyk, B. Lortz, Karl L. Nickel, and Hans-Wilm Wippermann. The algorithmic language Triplex-ALGOL 60. *Numer. Math.*, 11(2):175–180, Feb. 1968.

[19] Nikolaos Apostolatos, Ulrich W. Kulisch, and Karl L. Nickel. Ein Einschliessungsverfahren für Nullstellen. *Computing*, 2:195–201, 1967.

[20] Eckart Baumann. Optimal centered forms. *BIT*, 28(1):80–87, 1988.

[21] Martin Berz, Christian Bischof, George F. Corliss, and Andreas Griewank, editors. *Computational Differentiation: Techniques, Applications, and Tools*. SIAM, Philadelphia, 1996.

[22] Martin Berz and Kyoko Makino. Higher order multivariate automatic differentiation and validated computation of remainder bounds. *WSEAS Trans. Math.*, 3(1):37–44, Jan. 1998.

[23] Martin Berz and Kyoko Makino. New methods for high-dimensional verified quadrature. *Reliable Computing*, 5:13–22, Feb. 1999.

[24] Martin Berz and Kyoko Makino. Performance of Taylor model methods for validated integration of ODEs. In Jack Dongarra, Kaj Madsen, and Jerzy Wasniewski, editors, Applied Parallel Computing, *Lecture Notes in Computer Science* 3732, Springer-Verlag, New York, 2005, pp. 65–74.

[25] Martin Berz, Kyoko Makino, and Youn-Kyung Kim. Long-term stability of the tevatron by verified global optimization. *Nuclear Instruments Methods Physics Research A*, 558:1–10, March 2006.

[26] W. Binstock, J. Hawkes, and N.-T. Hsu. *An Interval Input / Output Package for the Univac 1108*. Technical Report 1212, Mathematics Research Center, University of Wisconsin, Sept. 1973.

[27] Folkmar Bornemann, Dirk Laurie, Stanley Wagon, and Jörg Waldvogel. *The SIAM 100-Digit Challenge: A Study in High-Accuracy Numerical Computing*. SIAM, Philadelphia, 2004.

[28] Glencora Borradaile and Pascal Van Hentenryck. Safe and tight linear estimators for global optimization. *Math. Program.*, 102(3):495–517, 2005.

[29] Norbert C. Börsken. *Komplexe Kreis-Standardfunktionen*. Dissertation, Universität Freiburg, Freiburg, Germany, 1978.

[30] Judy A. Braun and Ramon E. Moore. *A Program for the Solution of Differential Equations Using Interval Arithmetic (DIFEQ) for the CDC 3600 and 1604*. Technical Report 901, Mathematics Research Center, University of Wisconsin, Dec. 1967.

[31] Robert P. Broadwater, Hesham E. Shaalan, and Wolter J. Fabrycky. Decision evaluation with interval mathematics: A power distribution system case study. *IEEE Trans. Power Systems*, 20(2):725–7334, May 2005.

[32] Ole Caprani, Kaj Madsen, and Louis B. Rall. *Integration of Interval Functions. SIAM J. Math. Anal.*, vol 2: 321–341, 1981.

[33] Ole Caprani, Kaj Madsen, and Louis B. Rall. Integration of interval functions. *SIAM J. Math. Anal.*, 12(3):321–341, 1981.

[34] Damien Chablat, Philippe Wenger, Félix Majou, and Jean-Pierre Merlet. An interval analysis based study for the design and the comparison of three-degrees-of-freedom parallel kinematic machines. *I. J. Robotic Res.*, 23(6):615–624, 2004.

[35] Su Huan Chen, Hua Dong Lian, and Xiao Wei Yang. Interval eigenvalue analysis for structures with interval parameters. *Finite Elements Anal. Design*, 39:419–431, March 2003.

[36] Bruce Christianson. Reverse accumulation and accurate rounding error estimates for Taylor series coefficients. *Optimization Methods and Software*, 1(1):81–94, 1991.

[37] Amanda E. Connell and Robert. M. Corless. An experimental interval arithmetic package in Maple. *Interval Comput.*, 2:120–134, 1993.

[38] George F. Corliss, Christèle Faure, Andreas Griewank, Laurent Hascoët, and Uwe Naumann, editors. *Automatic Differentiation of Algorithms: From Simulation to Optimization*. Computer and Information Science. Springer, New York, 2001.

[39] George F. Corliss, Christopher Foley, and Ralph Baker Kearfott. Formulation for reliable analysis of structural frames. *Reliable Comput.*, 13:125–147, April 2007.

[40] George F. Corliss and Louis B. Rall. Adaptive, self-validating numerical quadrature. *SIAM J. Sci. Statist. Comput.*, 8(5):831–847, 1987.

[41] I. Csajka, W. Muenzer, and Karl L. Nickel. *Subroutines add, neg, sub, div, mul, for Use in an Error-Bound Arithmetic*. Technical report, IBM Research Laboratory, Rueschlikon, Switzerland, 1966.

[42] Brian Davies. Whither mathematics? *Notices Amer. Math. Soc.*, 52(11):1350–1356, Dec. 2005.

[43] Kaisheng Du. *Cluster Problem in Global Optimization Using Interval Arithmetic*. Ph.D. thesis, University of Southwestern Louisiana, 1994.

[44] Kaisheng Du and Ralph Baker Kearfott. The cluster problem in global optimization: The univariate case. *Computing (Suppl.)*, 9:117–127, 1992.

[45] Bo Einarsson, editor. *Accuracy and Reliability in Scientific Computing*. SIAM, Philadelphia, 2005.

[46] Andrea Fusiello, Arrigo Benedetti, Michela Farenzena, and Alessandro Busti. Globally convergent autocalibration using interval analysis. *IEEE Trans. Pattern Anal. Mach. Intell.*, 26(12):1633–1638, 2004.

[47] Irene Gargantini and Peter Henrici. Circular arithmetic and the determination of polynomial zeros. *Numer. Math.*, 18(4):305–320, Aug. 1972.

[48] Chao-Yang Gau and Mark A. Stadtherr. Dynamic load balancing for parallel interval-Newton using message passing. *Comput. Chemical Engrg.*, 26:811–815, June 2002.

[49] Chao-Yang Gau and Mark A. Stadtherr. New interval methodologies for reliable chemical process modeling. *Comput. Chemical Engrg.*, 26:827–840, June 2002.

[50] Gene H. Golub and Charles F. van Loan. *Matrix Computations*, 3rd ed. Johns Hopkins University Press, Baltimore, 1996.

[51] Laurent Granvilliers and Frédéric Benhamou. Progress in the solving of a circuit design problem. *J. Global Optim.*, 20:155–168, June 2001.

[52] Laurent Granvilliers and Frédéric Benhamou. Algorithm 852: RealPaver: An interval solver using constraint satisfaction techniques. *ACM Trans. Math. Softw.*, 32(1):138–156, 2006.

[53] Andreas Griewank. *Evaluating Derivatives: Principles and Techniques of Algorithmic Differentiation*. Frontiers in Appl. Math., 19, SIAM, Philadelphia, 2000.

[54] Andreas Griewank and George F. F. Corliss, editors. *Automatic Differentiation of Algorithms: Theory, Implementation, and Application*. SIAM, Philadelphia, 1991.

[55] Angelika C. Hailer. *Verification of Branch and Bound Algorithms Applied to Water Distribution Network Design*. Logos Verlag, Berlin, 2006. Dissertation, Technische Universität Hamburg, Harburg.

[56] Rolf Hammer, Matthias Hocks, Ulrich W. Kulisch, and Dietmar Ratz. *Numerical Toolbox for Verified Computing I—Basic Numerical Problems*. Springer-Verlag, Heidelberg, 1993.

[57] E. Hansen and G. W. Walster. *Global Optimization Using Interval Analysis*. Marcel Dekker, New York, 2003.

[58] Eldon R. Hansen. Interval arithmetic in matrix computations, part I. *J. Soc. Indust. Appl. Math. Ser. B Numer. Anal.*, 2(2):308–320, 1965.

[59] Eldon R. Hansen. On solving systems of equations using interval arithmetic. *Math. Comput.*, 22(102):374–384, April 1968.

[60] Eldon R. Hansen. The centered form. In Eldon R. Hansen, editor, *Topics in Interval Analysis*, Oxford University Press, London, 1969, pp. 102–106.

[61] Eldon R. Hansen. On linear algebraic equations with interval coefficients. In Eldon R. Hansen, editor, *Topics in Interval Analysis*, Oxford University Press, London, 1969, pp. 35–46.

[62] Eldon R. Hansen. On solving two-point boundary-value problems using interval arithmetic. In Eldon R. Hansen, editor, *Topics in Interval Analysis*, Oxford University Press, London, 1969, pp. 74–90.

[63] Eldon R. Hansen. *Topics in Interval Analysis*. Oxford University Press, London, 1969.

[64] Eldon R. Hansen. Interval forms of Newton's method. *Computing*, 20:153–163, 1978.

[65] Eldon R. Hansen. *Global Optimization Using Interval Analysis*. Marcel Dekker, New York, 1992.

[66] Eldon R. Hansen and Saumyendra Sengupta. Bounding solutions of systems of equations using interval analysis. *BIT (Nordisk tidskrift for informationsbehandling)*, 21(2):203–211, June 1981.

[67] Eldon R. Hansen and Roberta Smith. *A Computer Program for Solving a System of Linear Equations and Matrix Inversion with Automatic Error Bounding Using Interval Arithmetic*. Technical Report LMSC 4-22-66-3, Lockheed Missiles and Space Co., Palo Alto, CA, 1966.

[68] Eldon R. Hansen and Roberta Smith. Interval arithmetic in matrix computations, part II. *SIAM J. Numer. Anal.*, 4(1):1–9, March 1967.

[69] Eldon R. Hansen and G. William Walster. Global optimization in nonlinear mixed integer problems. In William F. Ames and R. Vichnevetsky, editors, *Proceedings of the 10th IMACS World Congress on Systems Simulation and Scientific Computing*, vol. 1, IMACS, Plantation, FL, 1982, pp. 379–381.

[70] G. I. Hargreaves. *Interval Analysis in MATLAB*. Master's thesis, Department of Mathematics, University of Manchester, 2002.

[71] Joel Hass, Michael Hutchings, and Roger Schlafly. The double bubble conjecture. *Electronic Research Announcements of the American Mathematical Society*, 1:98–102, 1995.

[72] Peter Henrici, editor. *Applied and Computational Complex Analysis*. Wiley, New York, 1977.

[73] Jürgen Herzberger. *Metrische Eigenschaften von Mengensystemen und einige Anwendungen*. Dissertation, Universität Karlsruhe, Germany, 1969.

[74] Jürgen Herzberger. Intervallmässige Auswertung von Standardfunktionen in Algol-60. *Computing*, 5:377–384, 1970.

[75] Jürgen Herzberger. Zur Konvergenz intervallmässiger Iterationsverfahren. *Z. Angew. Math. Mech.*, 51:T56, 1971.

[76] Jürgen Herzberger. Über ein Verfahren zur Bestimmung reeller Nullstellen mit Anwendung auf Parallelrechnung. *Elektronische Rechenanlagen*, 14:250–254, 1972.

[77] Jürgen Herzberger. *Über optimale Verfahren zur numerischen Bestimmung reeller Nullstellen mit Fehlerschranken*. Habilitationsschrift, Universität Karlsruhe, Germany, 1972.

[78] Jürgen Herzberger. Bemerkungen zu einem Verfahren von Ramon E. Moore. *Z. Angew. Math. Mech.*, 53:356–358, 1973.

[79] Desmond J. Higham and Nicholas J. Higham. *MATLAB Guide*. SIAM, Philadelphia, 2000.

[80] Werner Hofschuster and Walter Krämer. C-XSC 2.0—A C++ library for extended scientific computing. *Lecture Notes in Comput. Sci.*, 2991, Springer-Verlag, New York, 2004, pp. 15–35.

[81] Oliver Holzmann, Bruno Lang, and Holger Schütt. Newton's constant of gravitation and verified numerical quadratures. *Reliable Comput.*, 2(3):229–240, 1996.

[82] Chenyi Hu and Ralph Baker Kearfott. On bounding the range of some elementary functions in FORTRAN 77. *Interval Comput.*, 1993(3):29–39, 1993.

[83] Umran S. Inan and Aziz S. Inan. *Electromagnetic Waves*. Prentice–Hall, Upper Saddle River, NJ, 2000.

[84] Christian Jansson. Rigorous lower and upper bounds in linear programming. *SIAM J. Optim.*, 14(3):914–935, 2004.

[85] Christian Jansson, Denis Chaykin, and Christian Keil. Rigorous error bounds for the optimal value in semidefinite programming. *SIAM J. Numer. Anal.*, 46(1):180–200, 2007.

[86] Luc Jaulin, Michel Keiffer, Olivier Didrit, and Eric Walter. *Applied Interval Analysis*. Springer-Verlag, Berlin, 2001.

[87] William M. Kahan. A computable error-bound for systems of ordinary differential equations (abstract). *SIAM Rev.*, 8:568–569, 1966.

[88] William M. Kahan. Circumscribing an ellipsoid about the intersection of two ellipsoids. *Can. Math. Bull.*, 11:437–441, 1968.

[89] William M. Kahan. *A More Complete Interval Arithmetic: Lecture Notes for an Engineering Summer Course in Numerical Analysis at the University of Michigan.* Technical report, University of Michigan, 1968.

[90] William M. Kahan. Invited commentary (concerning the invited paper of Karl L. Nickel: Error bounds and computer arithmetic). In A. Morrell, editor, *Proceedings of IFIP Congress 1968*, vol. I, North–Holland, Amsterdam, 1969, pp. 60–62.

[91] William M. Kahan. A survey of error analysis. In C. V. Freiman, J. E. Griffith, and J. L. Rosenfeld, editors, *Proceedings of IFIP Congress 1971*, vol. II, North–Holland, Amsterdam, 1972, pp. 1214–1239.

[92] Leonid Vitalevich Kantorovich. On a mathematical symbolism convenient for performing machine calculations (in Russian). *Dokl. Akad. Nauk SSSR*, 113(4):738–741, 1957.

[93] Ralph Baker Kearfott. Abstract generalized bisection and a cost bound. *Math. Comput.*, 49(179):187–202, July 1987.

[94] Ralph Baker Kearfott. Some tests of generalized bisection. *ACM Trans. Math. Software*, 13(3):197–220, Sept. 1987.

[95] Ralph Baker Kearfott. Corrigenda: Some Tests of Generalized Bisection. *ACM Trans. Math. Software*, 14(4):399–399, Dec. 1988.

[96] Ralph Baker Kearfott. Algorithm 763: INTERVAL_ARITHMETIC: A Fortran 90 module for an interval data type. *ACM Trans. Math. Software*, 22(4):385–392, Dec. 1996.

[97] Ralph Baker Kearfott. *Rigorous Global Search: Continuous Problems.* Nonconvex Optimization and Its Applications 13, Kluwer Academic Publishers, Dordrecht, The Netherlands, 1996.

[98] Ralph Baker Kearfott. On proving existence of feasible points in equality constrained optimization problems. *Math. Program.*, 83(1):89–100, 1998.

[99] Ralph Baker Kearfott. GlobSol User Guide, 2008. To appear in *Optim. Methods Software.*

[100] Ralph Baker Kearfott. Verified branch and bound for singular linear and nonlinear programs: An epsilon-inflation process, April 2007. Available from the author.

[101] Ralph Baker Kearfott, Milinde Dawande, Kaisheng Du, and Chenyi Hu. Algorithm 737: INTLIB: A portable Fortran-77 elementary function library. *ACM Trans. Math. Software*, 20(4):447–459, Dec. 1994.

[102] Ralph Baker Kearfott and Kaisheng Du. The cluster problem in multivariate global optimization. *J. Global Optim.*, 5:253–265, 1994.

[103] Ralph Baker Kearfott and Vladik Kreinovich, editors. Applications of Interval Computations: Papers Presented at an International Workshop in El Paso, Texas, February 23–25, 1995, *Applied Optimization* 3, Kluwer, Dordrecht, The Netherlands, 1996.

[104] Ralph Baker Kearfott, Markus Neher, Shin'ichi Oishi, and Fabien Rico. Libraries, tools, and interactive systems for verified computations: Four case studies. *Lecture Notes in Comput. Sci.*, 2991, Springer-Verlag, New York, 2004, pp. 36–63.

[105] Ralph Baker Kearfott and Manuel Novoa III. Algorithm 681: INTBIS, a portable interval Newton/bisection package. *ACM Trans. Math. Software*, 16(2):152–157, June 1990.

[106] Ralph Baker Kearfott and G. William Walster. On stopping criteria in verified nonlinear systems or optimization algorithms. *ACM Trans. Math. Software*, 26(3):373–389, Sept. 2000.

[107] Gershon Kedem. A posteriori error bounds for two-point boundary value problems. *SIAM J. Numer. Anal.*, 18(3):431–448, June 1981.

[108] Christian Keil and Christian Jansson. Computational experience with rigorous error bounds for the netlib linear programming library. *Reliable Comput.*, 12(4):303–321, 2006.

[109] J. B. Keiper. Interval arithmetic in Mathematica. *Interval Comput.*, 1993(3):76–87, 1993.

[110] Hassan K. Khalil. *Nonlinear Systems*. MacMillan, New York, 1992.

[111] Olaf Knüppel. PROFIL/BIAS—A fast interval library. *Computing*, 53(3–4):277–287, Sept. 1994.

[112] Olaf Knüppel. *PROFIL/BIAS v 2.0. Bericht 99.1*, Technische Universität Hamburg-Harburg, Harburg, Germany, Feb. 1999.

[113] Martin Koeber. Inclusion of solutions of Darboux problems for quasilinear hyperbolic equations. *J. Comput. Appl. Math.*, 152:243–262, March 2003.

[114] Lubo Kolev. *Interval Methods for Circuit Analysis*. World Scientific, Singapore, 2003.

[115] Hasan Ugur Köylüoglu, Ahmet Ş. Çakmak, and Søren R. K. Neilson. Interval algebra to deal with pattern loading and structural uncertainty. *J. Engrg. Mech.*, 121:1149–1157, 1995.

[116] Walter Krämer. intpakX—An Interval Arithmetic Package for Maple. In *Proceedings of the 12th GAMM, IMACS International Symposium on Scientific Computing, Computer Arithmetic and Validated Numerics (SCAN 2006)*, IEEE Computer Society, Washington, DC, 2006, p. 26.

[117] Walter Krämer, Ulrich W. Kulisch, and Rudolf Lohner. *Numerical Toolbox for Verified Computing II: Advanced Numerical Problems*. Springer-Verlag, New York, 2006.

[118] Walter Krämer and Stefan Wedner. Two adaptive Gauss–Legendre type algorithms for the verified computation of definite integrals. *Reliable Comput.*, 2(3):241–254, 1996.

[119] Rudolf Krawczyk. Newton-Algorithmen zur Bestimmung von Nullstellen mit Fehlerschranken. Interner Bericht des Inst. für Informatik 68/6, Universität Karlsruhe, 1968. *Computing*, 4:187–201, 1969.

[120] Rudolf Krawczyk and Arnold Neumaier. Interval slopes for rational functions and associated centered forms. *SIAM J. Numer. Anal.*, 22(3):604–616, June 1985.

[121] Vladik Kreinovich, Anatoly Lakeyev, Jiři Rohn, and Patrick Kahl. Computational Complexity and Feasibility of Data Processing and Interval Computations, *Applied Optimization* 10, Kluwer Academic Publishers, Dordrecht, The Netherlands, 1998.

[122] Ulrich W. Kulisch. An axiomatic approach to rounded computations. *Numer. Math.*, 18:1–17, 1971.

[123] Ulrich W. Kulisch. Grundzüge der Intervallrechnung. In Detlef Laugwitz, editor, *Überblicke Mathematik 2*, Bibliographisches Institut, Mannheim, 1969, pp. 51–98.

[124] Ulrich W. Kulisch. *Interval Arithmetic Over Completely Ordered Ringoids*. Technical Report 1105, Mathematics Research Center, University of Wisconsin, May 1972.

[125] Ulrich W. Kulisch and Willard L. Miranker. *Computer Arithmetic in Theory and Practice*. Academic Press, New York, 1981.

[126] Ulrich W. Kulisch and Willard L. Miranker. The arithmetic of the digital computer: A new approach. *SIAM Rev.*, 28(1):1–40, March 1986.

[127] Thomas D. Ladner and J. Michael Yohe. *An Interval Arithmetic Package for the Univac 1108*. Technical Report 1055, Mathematics Research Center, University of Wisconsin, Aug. 1970.

[128] Bruno Lang. Verified quadrature in determining Newton's constant of gravitation. *J. Universal Comput. Sci.*, 4(1):16–24, Jan. 1998.

[129] Bruno Lang. Derivative-based subdivision in multi-dimensional verified Gaussian quadrature. In Götz Alefeld, Jiri Rohn, Siegfried M. Rump, and Tetsuro Yamamoto, editors, *Symbolic Algebraic Methods and Verification Methods*, Springer-Verlag, New York, 2001, pp. 145–152.

[130] Yahia Lebbah, Claude Michel, Michel Rueher, David Daney, and Jean-Pierre Merlet. Efficient and safe global constraints for handling numerical constraint systems. *SIAM J. Numer. Anal.*, 42(5):2076–2097, 2005.

[131] Eric Lee, Constantinos Mavroidis, and Jean-Pierre Merlet. Five precision point synthesis of spatial RRR manipulators using interval analysis. *J. Mech. Design*, 126:842–849, 2004.

[132] Michael Lerch, German Tischler, Jürgen Wolff von Gudenberg, Werner Hofschuster, and Walter Krämer. FILIB++, a fast interval library supporting containment computations. *ACM Trans. Math. Software*, 32(2):299–324, June 2006.

[133] Youdong Lin, Joshua A. Enszer, and Mark A. Stadtherr. Enclosing all solutions of two-point boundary value problems for ODEs. *Comput. Chemical Engrg.*, 32(8):1714–1725, August 2008.

[134] Youdong Lin and Mark A. Stadtherr. Advances in interval methods for deterministic global optimization in chemical engineering. *J. Global Optim.*, 29:281–296, July 2004.

[135] Youdong Lin and Mark A. Stadtherr. Locating stationary points of sorbate-zeolite potential energy surfaces using interval analysis. *J. Chem. Phys.*, 121:10159–10166, 2004.

[136] Youdong Lin and Mark A. Stadtherr. LP strategy for interval-Newton method in deterministic global optimization. *Ind. Engrg. Chem. Res.*, 43:3741–3749, 2004.

[137] Youdong Lin and Mark A. Stadtherr. Deterministic global optimization of molecular structures using interval analysis. *J. Comput. Chem.*, 26:1413–1420, 2005.

[138] Kyoko Makino and Martin Berz. Taylor models and other validated functional inclusion methods. *Internat. J. Pure Appl. Math.*, 4,4:379–456, 2003.

[139] Kyoko Makino and Martin Berz. Suppression of the wrapping effect by Taylor model-based verified integrators: Long-term stabilization by preconditioning. *Int. J. Differ. Equ. Appl.*, 10(4):353–384, 2005.

[140] Günter Mayer. Direct methods for linear systems with inexact input data, 2008. To appear in *Japan J. Industrial Appl. Math.*

[141] Keith Meintjes and Alexander P. Morgan. *A Methodology for Solving Chemical Equilibrium Systems.* General Motors Research Report GMR-4971. General Motors Research Laboratories, Warren, MI, 1985.

[142] Jean-Pierre Merlet. ALIAS: An interval analysis based library for solving and analyzing system of equations. In *SEA*, Toulouse, June 2000.

[143] Kenneth R. Meyer and Dieter S. Schmidt, editors. *Computer Aided Proofs in Analysis.* Springer-Verlag, New York, 1991.

[144] Mehdi Modares, Robert L. Mullen, Paul X. Bellini, and R. Muhanna. Buckling analysis of structures with interval uncertainties. *SAE Transactions Journal of Passenger Cars—Mechanical Systems*, March 1996, pp. 284–290.

[145] Cleve B. Moler. *Numerical Computing with MATLAB.* SIAM, Philadelphia, 2004.

[146] Ramon E. Moore. *Interval Arithmetic and Automatic Error Analysis in Digital Computing.* Ph.D. dissertation, Department of Mathematics, Stanford University, Stanford, CA, 1962.

[147] Ramon E. Moore. The automatic analysis and control of error in digital computing based on the use of interval numbers. In Louis B. Rall, editor, *Error in Digital Computation*, vol. I, Wiley, New York, 1965, pp. 61–130.

[148] Ramon E. Moore. *Interval Analysis*. Prentice-Hall, Englewood Cliffs, NJ, 1966.

[149] Ramon E. Moore. *Mathematical Elements of Scientific Computing*. Holt, Rinehart, and Winston, New York, 1975.

[150] Ramon E. Moore. A test for existence of solutions to nonlinear systems. *SIAM J. Numer. Anal.*, 14(4):611–615, 1977.

[151] Ramon E. Moore. A computational test for convergence of iterative methods for nonlinear systems. *SIAM J. Numer. Anal.*, 15(6):1194–1196, 1978.

[152] Ramon E. Moore. *Methods and Applications of Interval Analysis*. SIAM, Stud. Appl. Math. 2, Philadelphia, 1979.

[153] Ramon E. Moore. Parameter sets for bounded error data. *Math. Comput. Simulation*, 34:113–119, 1992.

[154] Ramon E. Moore. The resolution of close minima. *Comput. Math. Appl.*, 25:57–58, 1993.

[155] Ramon E. Moore. The dawning. *Reliable Comput.*, 5:423–424, 1999.

[156] Ramon E. Moore and Michael J. Cloud. *Computational Functional Analysis, 2nd ed.* Horwood Publishing, London, 2006.

[157] Ramon E. Moore, J. Ann Davison, H. Riley Jaschke, and Sidney Shayer. *DIFEQ Integration Routine—User's Manual.* Technical Report LMSC6-90-64-6, Lockheed Missiles and Space, Lockheed, CA, 1964.

[158] Ramon E. Moore, Eldon R. Hansen, and Anthony Leclerc. Rigorous methods for parallel global optimization. In A. Floudas and P. Pardalos, editors, *Recent Advances in Global Optimization*, Princeton University Press, Princeton, NJ, 1992, pp. 321–342.

[159] Ramon E. Moore and Sandie T. Jones. Safe starting regions for iterative methods. *SIAM J. Numer. Anal.*, 14(6):1051–1065, 1977.

[160] Rafi L. Muhanna and Robert L. Mullen. Development of interval based methods for fuzziness in continuum-mechanics. In *ISUMA '95: Proceedings of the 3rd International Symposium on Uncertainty Modelling and Analysis*, IEEE Computer Society, Washington, DC, 1995, p. 705.

[161] Rafi L. Muhanna and Robert L. Mullen. Formulation of fuzzy finite-element methods for mechanics problems. *Computer-Aided Civil and Infrastructure Engineering*, 14(2):107–117, 1999.

[162] Rafi L. Muhanna and Robert L. Mullen. Uncertainty in mechanics problems. *J. Engrg. Mech.*, 127:557–566, 2001.

[163] Rafi L. Muhanna, Hao Zhang, and Robert L. Mullen. Interval finite elements as a basis for generalized models of uncertainty in engineering mechanics. *Reliable Comput.*, 13:173–194, April 2007.

[164] Kaori Nagatou. A computer-assisted proof on the stability of the Kolmogorov flows of incompressible viscous fluid. *J. Comput. Appl. Math.*, 169(1):33–44, 2004.

[165] Yusuke Naka, Assad A. Oberai, and Barbara G. Shinn-Cunningham. Acoustic eigenvalues of rectangular rooms with arbitrary wall impedances using the interval Newton / generalized bisection method. *J. Acoust. Soc. Am.*, 118:3662–3771, December 2005.

[166] Mitsuhiro T. Nakao. Numerical verification methods for solutions of ordinary and partial differential equations. *Numer. Funct. Anal. Optim.*, 22:321–356, 2001.

[167] Arnold Neumaier. Interval Methods for Systems of Equations. *Encyclopedia of Mathematics and its Applications* 37, Cambridge University Press, Cambridge, UK, 1990.

[168] Arnold Neumaier. A simple derivation of the Hansen-Bliek-Rohn-Ning-Kearfott enclosure for linear interval equations. *Reliable Comput.*, 5:131–136, 1999.

[169] Arnold Neumaier. Taylor forms—use and limits. *Reliable Comput.*, 9(1):43–79, 2003.

[170] Arnold Neumaier. Complete search in continuous global optimization and constraint satisfaction. In A. Iserles, editor, *Acta Numerica 2004*, Cambridge University Press, Cambridge, UK, 2004, pp. 271–369.

[171] Arnold Neumaier, Christian Bliek, and Christophe Jermann. Global Optimization and Constraint Satisfaction: First International Workshop Global Constraint Optimization and Constraint Satisfaction, Cocos 2002, Valbonne-Sophia Antipolis, France, October 2002. *Lecture Notes in Computer Science* 2861, Springer-Verlag, New York, 2004.

[172] Arnold Neumaier and Andrzej Pownuk. Linear systems with large uncertainties, with applications to truss structures. *Reliable Comput.*, 13:149–172, April 2007.

[173] Arnold Neumaier and Oleg Shcherbina. Safe bounds in linear and mixed-integer programming. *Math. Prog.*, 99(2):283–296, March 2004.

[174] Arnold Neumaier, Oleg Shcherbina, Waltraud Huyer, and Tamás Vinkó. A comparison of complete global optimization solvers. *Math. Prog.*, 103(2):335–356, 2005.

[175] Karl L. Nickel. Die numerische Berechnung der Wurzeln eines Polynoms. *Numer. Math.*, 9(1):80–98, Nov. 1966.

[176] Karl L. Nickel. Über die Notwendigkeit einer Fehlerschranken- Arithmetik für Rechenautomaten. *Numer. Math.*, 9:69–79, 1966.

[177] Karl L. Nickel. Die vollautomatische Berechnung einer einfachen Nullstelle von $F(T) = 0$ einschließlich einer Fehlerabschätzung. *Computing*, 2:232–245, 1967.

[178] Karl L. Nickel. Quadraturverfahren mit Fehlerschranken. *Z. Angew. Math. Mech.*, 47:T68–T69, 1967.

[179] Karl L. Nickel. Zwei neue Rechenmaschinen-Systeme an der TH Karlsruhe: Hydra-X8 und Triplex-Algol-Z23. *Umschau*, 67:525–526, 1967.

[180] Karl L. Nickel. *On the Newton Method in Interval Analysis*. Technical Report 1136, Mathematics Research Center, University of Wisconsin, Dec. 1971.

[181] Karl L. Nickel. *The Contraction Mapping Fixpoint Theorem in Interval Analysis*. Technical Report 1334, Mathematics Research Center, University of Wisconsin, March 1973.

[182] Karl L. Nickel. The over-estimation of the range of a function in interval mathematics with applications to the solution of linear systems of equations. Technical Report 1593, Mathematics Research Center, University of Wisconsin, Dec. 1975. Appeared as Die Überschätzung des Wertebereichs einer Funktion in der Intervallrechnung mit Anwendungen auf lineare Gleichungssysteme, *Computing*, 18(1):15–36, 1977. (March, 1977).

[183] Karl L. Nickel. Using interval methods for the numerical solution of ODE's. *Z. Angew. Math. Mech.*, 66:513–523, 1986.

[184] Shiying Ning and Ralph Baker Kearfott. A comparison of some methods for solving linear interval equations. *SIAM J. Numer. Anal.*, 34(4):1289–1305, 1997.

[185] Takeshi Ogita, Siegfried M. Rump, and Shin'ichi Oishi. Accurate sum and dot product. *SIAM J. Sci. Comput.*, 26(6):1955–1988, 2005.

[186] Shin'ichi Oishi, Kunio Tanabe, Takeshi Ogita, and Siegfried M. Rump. Convergence of Rump's method for inverting arbitrarily ill-conditioned matrices. *J. Comput. Appl. Math.*, 205(1):533–544, 2007.

[187] F. Aleixo Oliveira. Interval analysis and two-point boundary value problems. *SIAM J. Numer. Anal.*, 11(2):382–391, 1974.

[188] James M. Ortega and Werner C. Rheinboldt. *Iterative Solution of Nonlinear Equations in Several Variables*. Academic Press, New York, 1970.

[189] Katsuhisa Ozaki, Takeshi Ogita, Shinya Miyajima, Shin'ichi Oishi, and Siegfried M. Rump. A method of obtaining verified solutions for linear systems suited for Java. *J. Comput. Appl. Math.*, 199(2):337–344, 2007.

[190] Miodrag S. Petković and Ljiljana D. Petković. *Complex Interval Arithmetic and Its Applications*. Wiley, New York, 1998.

[191] Knut Petras. Principles of verified numerical integration. *J. Comput. Appl. Math.*, 199(2):317–328, 2007.

[192] Aurelio Piazzi and Antonio Visioli. Global minimum-jerk trajectory planning of robot manipulators. *IEEE Trans. Industrial Electronics*, 47:140–149, Feb. 2000.

[193] Michael Plum. Computer-assisted enclosure methods for elliptic differential equations. *Linear Algebra and Its Applications*, 324:147–187, Feb. 2001.

[194] John Derwent Pryce and George F. Corliss. Interval arithmetic with containment sets. *Computing*, 78(3):251–276, 2006.

[195] Louis B. Rall, editor. *Error in Digital Computation*, vol. I. Wiley, New York, 1965.

[196] Louis B. Rall, editor. *Error in Digital Computation*, vol. II. Wiley, New York, 1965.

[197] Louis B. Rall. Application of interval integration to the solution of integral equations. *J. Integral Equations*, 6:127–141, 1984.

[198] Louis B. Rall. Representations of intervals and optimal error bounds. *Math. Comput.*, 41:219–227, 1983.

[199] Louis B. Rall. Application of interval integration to the solution of integral equations II—the finite case. *SIAM J. Math. Anal.*, 13:690–697, 1982.

[200] Louis B. Rall. *Automatic Differentiation:* Techniques and Applications. *Lecture Notes in Computer Science 120.* Springer-Verlag, Berlin, 1981.

[201] Louis B. Rall. Interval analysis: A new tool for applied mathematics. In *Transactions of the Twenty-Seventh Conference of Army Mathematicians*, Army Research Office, Research Triangle Park, NC, 1982, pp. 283–301.

[202] Louis B. Rall. Mean value and Taylor forms in interval analysis. *SIAM J. Math. Anal.*, 14:223–238, 1983.

[203] Louis B. Rall. *Solution of Finite Systems of Equations by Interval Iteration.* Technical Report 2271, Mathematics Research Center, University of Wisconsin, Aug. 1981.

[204] Louis B. Rall. A theory of interval iteration. *Proceedings of the American Mathematical Society*, 86:625–631, 1982.

[205] Louis B. Rall. Interval bounds for stationary values of functionals. *Nonlinear Anal.*, 6:855–861, 1982.

[206] Louis B. Rall. Interval methods for fixed-point problems. *Numerical Functional Analysis and Optimization*, 9:35–59, 1987.

[207] Louis B. Rall. *Global Optimization Using Automatic Differentiation and Interval Iteration.* Technical Report 2832, Mathematics Research Center, University of Wisconsin, June 1985.

[208] Louis B. Rall. Improved interval bounds for ranges of functions. *Lecture Notes in Comput. Sci.* 212, Karl L. Nickel, editor, Springer-Verlag, New York, 1986, pp. 143–155.

[209] Helmut Ratschek. Centered forms. *SIAM J. Numer. Anal.*, 17(5):656–662, 1980.

[210] Helmut Ratschek and Jon G. Rokne. *Computer Methods for the Range of Functions. Ellis Horwood Ser.: Math. Appl.* Ellis Horwood, Chichester, UK, 1984.

[211] Helmut Ratschek and Jon G. Rokne. *New Computer Methods for Global Optimization.* Wiley, New York, 1988.

[212] Helmut Ratschek and Jon G. Rokne. Experiments using interval analysis for solving a circuit design problem. *J. Global Optim.*, 3:501–518, Dec. 1993.

[213] Dietmar Ratz. *Automatic Slope Computation and Its Application in Nonsmooth Global Optimization.* Shaker Verlag, Aachen, 1998.

[214] Gerald Recktenwald. *Numerical Methods with MATLAB: Implementations and Applications.* Prentice–Hall, Englewood Cliffs, NJ, 2000.

[215] Allen Reiter. *Interval Arithmetic Package (INTERVAL) for the CDC 1604 and CDC 3600.* Technical Report 794, Mathematics Research Center, University of Wisconsin, Dec. 1967.

[216] Thomas W. Reps and Louis B. Rall. Computational divided differencing and divided-difference arithmetics. *Higher Order Symbol. Comput.*, 16(1-2):93–149, 2003.

[217] Nathalie Revol and Fabrice Rouillier. Motivations for an arbitrary precision interval arithmetic and the MPFI library. *Reliable Comput.*, 11(4):275–290, 2005.

[218] Siegfried M. Rump. *Kleine Fehlerschranken bei Matrixproblemen.* Dissertation, Universität Karlsruhe, Germany, 1980.

[219] Siegfried M. Rump. Solving algebraic systems with high accuracy. In Ulrich W. Kulisch and Willard L. Miranker, editors, *A New Approach to Scientific Computation*, Academic Press, New York, 1983, pp. 51–120.

[220] Siegfried M. Rump. Algebraic Computation, Numerical Computation and Verified Inclusions. In R. Janssen, editor, Trends in Computer Algebra, *Lecture Notes in Computer Science* 296, 1988, pp. 177–197.

[221] Siegfried M. Rump. Verification methods for dense and sparse systems of equations. In Jürgen Herzberger, editor, Topics in Validated Computations: Proceedings of IMACS-GAMM International Workshop on Validated Computation, Oldenburg, Germany, 30 August–3 September 1993, *Studies in Computational Mathematics* 5, Elsevier, Amsterdam, The Netherlands, 1994, pp. 63–136.

[222] Siegfried M. Rump. Expansion and estimation of the range of nonlinear functions. *Math. Comput.*, 65(216):1503–1512, Oct. 1996.

[223] Siegfried M. Rump. Fast and parallel interval arithmetic. *BIT Numer. Math.*, 39(3):534–554, Sep. 1999.

[224] Siegfried M. Rump. INTLAB–INTerval LABoratory. In Tibor Csendes, editor, Developments in Reliable Computing: Papers presented at the International Symposium on Scientific Computing, Computer Arithmetic, and Validated Numerics, SCAN-98, in Szeged, Hungary, *Reliable Computing* 5(3), Kluwer Academic Publishers, Dordrecht, The Netherlands, 1999, pp. 77–104.

[225] Siegfried M. Rump. Rigorous and portable standard functions. *BIT Numer. Math.*, 41(3):540–562, June 2001.

[226] Siegfried M. Rump and Takeshi Ogita. Super-fast validated solution of linear systems. *J. Comput. Appl. Math.*, 199(2):199–206, 2007.

[227] Nikolaos V. Sahinidis. BARON: A general purpose global optimization software package. *J. Global Optim.*, 8(2):201–205, 1996.

[228] Hermann Schichl and Arnold Neumaier. Exclusion regions for systems of equations. *SIAM J. Numer. Anal.*, 42(1):383–408, 2004.

[229] Marco Schnurr. *Steigungen höherer Ordnung zur verifizierten globalen Optimierung.* Ph.D. dissertation, Department of Mathematics, University of Karlsruhe, Germany, May 2007.

[230] Hesham E. Shaalan and Robert P. Broadwater. Using interval mathematics in cost benefit analysis of distribution automation. *Electric Power Systems Research*, 27:145–152, 1993.

[231] Zhixin Shi and Brian Hassard. Precise solution of Laplace's equation. *Math. Comput.*, 64(210):515–536, April 1995.

[232] Stig Skelboe. Computation of rational interval functions. *BIT*, 14:187–195, 1974.

[233] John M. Snyder. *Generative Modeling for Computer Graphics and CAD: Symbolic Shape Design Using Interval Analysis.* Academic Press Professional, San Diego, CA, 1992.

[234] Ivan S. Sokolnikoff and Raymond M. Redheffer. *Mathematics of Physics and Modern Engineering.* McGraw-Hill, New York, 1966.

[235] Bert Speelpenning. *Compiling Fast Partial Derivatives of Functions Given by Algorithms.* Ph.D. thesis, Department of Computer Science, University of Illinois, Urbana-Champaign, Jan. 1980.

[236] Mark A. Stadtherr. Interval analysis: Application to chemical engineering design problems. In Arieh Iserles, editor, *Encyclopedia of Optimization*, Kluwer Academic Publishers, Dordrecht, The Netherlands, 2001.

[237] Mark A. Stadtherr. Interval analysis: Application to phase equilibrium problems. In Arieh Iserles, editor, *Encyclopedia of Optimization*. Kluwer Academic Publishers, Dordrecht, The Netherlands, 2001.

[238] Teruo Sunaga. Theory of interval algebra and its application to numerical analysis. *RAAG Memoirs*, 2:29–46, 1958.

[239] Peter Gabor Szabó, Mihaly Csaba. Markót, Tibor Csendes, Eckard Specht, Leoncadio G. Casado, and Inmaculada García. *New Approaches to Circle Packing in a Square: With Program Codes, Springer Optimization and Its Applications*. Springer-Verlag, New York, 2007.

[240] George G. Szpiro. *Kepler's Conjecture: How Some of the Greatest Minds in History Helped Solve One of the Oldest Math Problems in the World*. Wiley, New York, 2003.

[241] Mohit Tawarmalani and Nikolaos V. Sahinidis. *Convexification and Global Optimization in Continuous and Mixed-Integer Nonlinear Programming: Theory, Algorithms, Software, and Applications*. Kluwer Academic Publishers, Dordrecht, The Netherlands, 2002.

[242] Kenji Toyonaga, Mitsuhiro T. Nakao, and Yoshitaka Watanabe. Verified numerical computations for multiple and nearly multiple eigenvalues of elliptic operators. *J. Comput. Appl. Math.*, 147(1):175–190, 2002.

[243] Lung-Wen Tsai and Alexander P. Morgan. Solving the kinematics of the most general six- and five-degree-of-freedom manipulators by continuation methods. *ASME J. Mechanisms Transmissions Automation in Design*, 107:189–200, 1985.

[244] Warwick Tucker. A rigorous ODE solver and Smale's 14th problem. *Found. Comput. Math.*, 24:53–117, 2002.

[245] Pascal Van Hentenryck, Laurent Michel, and Yves Deville. *Numerica: A Modeling Language for Global Optimization*. MIT Press, Cambridge, MA, 1997.

[246] John von Neumann and Herman Heine Goldstine. Numerical inversion of matrices of high order. *Bull. Amer. Math. Soc.*, 53:1021–1099, 1947.

[247] G. William Walster, Eldon R. Hansen, and Saumyendra Sengupta. Test results for a global optimization algorithm. In Paul T. Boggs, Richard H. Byrd, and Robert B. Schnabel, editors, *Numerical Optimization 1984*, SIAM, Philadelphia, 1985, pp. 272–287.

[248] James Hardy Wilkinson. *The Algebraic Eigenvalue Problem*. Clarendon Press, Oxford, UK, 1988.

[249] Stephen J. Wright. A collection of problems for which Gaussian elimination with partial pivoting is unstable. *SIAM J. Sci. Comput.*, 14(1):231–238, 1993.

[250] Nobito Yamamoto. A numerical verification method for solutions of boundary value problems with local uniqueness by Banach's fixed-point theorem. *SIAM J. Numer. Anal.*, 35(5):2004–2013, 1998.

[251] J. Michael Yohe. Rigorous bounds on computed approximation to square roots and cube roots. *Computing*, 11:51–57, 1973.

[252] J. Michael Yohe. *Application of Interval Analysis to Error Control.* Technical Report 1686, Mathematics Research Center, University of Wisconsin, Sep. 1976.

[253] J. Michael Yohe. *The Interval Arithmetic Package.* Technical Report 1755, Mathematics Research Center, University of Wisconsin, June 1977.

[254] J. Michael Yohe. Software for interval arithmetic: A reasonably portable package. *ACM Trans. Math. Software*, 5:50–65, 1979.

[255] J. Michael Yohe. *The Interval Arithmetic Package—Multiple Precision Version.* Technical Report 1908, Mathematics Research Center, University of Wisconsin, Jan. 1979.

[256] Jiyuan. Zhang and Shoufan Shen. Interval analysis method for determining kinematic solutions of mechanisms. *Chinese J. Mech. Engrg.*, 27:75–99, 1991.

[257] Shen Zuhe and Michael A. Wolfe. On interval enclosures using slope arithmetic. *Appl. Math. Comput.*, 39(1):89–105, 1990.

Index